Meaningful Technologies

HOW DIGITAL METAPHORS CHANGE THE WAY WE THINK AND LIVE

Eric Chown and Fernando Nascimento

LEVER
PRESS

Lever Press (leverpress.org) is a publisher of pathbreaking scholarship. Supported by a consortium of higher education institutions focused on, and renowned for, excellence in both research and teaching, our press is grounded on three essential commitments: to be a digitally native press, to be a peer-reviewed, open access press that charges no fees to either authors or their institutions, and to be a press aligned with the liberal arts ethos.

DOI: https://doi.org/10.3998/mpub.12682261
PrintISBN:978- 1-64315-041-3
Open access ISBN: 978-1-64315-042-0

Library of Congress Control Number: 2022947923

Published in the United States of America by Lever Press, in partnership with Michigan Publishing.

Contents

Member Institution Acknowledgments

Lever Press is a joint venture. This work was made possible by the generous support of Lever Press member libraries from the following institutions:

Amherst College

Berea College

Bowdoin College

Carleton College

Central Washington University

Claremont Graduate University

Claremont McKenna College

Clark Atlanta University

College of Saint Benedict & Saint John's University

The College of Wooster

Davidson College

Denison University

DePauw University

Grinnell College

Hamilton College

Harvey Mudd College

Hollins University

Iowa State University

Keck Graduate Institute

Knox College

Lafayette College

Macalester College

Middlebury College

Morehouse College

Norwich University

Penn State University

Pitzer College

Pomona College

Randolph-Macon College

Rollins College

Santa Clara University

Scripps College

Skidmore College

Smith College

Spelman College

Susquehanna University

Swarthmore College
Trinity University
UCLA Library
Union College
University of Idaho
University of Northern
　Colorado
University of Puget Sound
University of San Francisco
University of Vermont

Ursinus College
Vassar College
Washington and Lee
　University
Whitman College
Whittier College
Whitworth University
Willamette University
Williams College

Acknowledgments

To my wife Rachel, who inspires me every day, and my kids, Kira and Zander, the best family I could ever hope for. To Crystal Hall and Pamela Fletcher for helping make DCS real. And to Fernando, who somehow has managed to make writing a book joyful.– Eric Chown

To Vanessa, Mateus, Filipe, and Beatriz, springs of meaning to my life. To Crystal Hall and George Taylor for their support, solicitude, and friendship. To Eric, an outstanding mentor and fantastic companion in this book journey.– Fernando Nascimento

INTRODUCTION

When it comes to technology, we as a society are obsessed with what is new, what it can do, and who made it. The press is full of stories about the latest gadgets and larger-than-life figures of the tech industry, such as Steve Jobs, Bill Gates, and Elon Musk. Look a bit further and you will find worries about tech's impact on individuals—digital addiction, digital depression, the digital divide, etc.—or on society, especially with regard to artificial intelligence. These are all important issues, and they are all connected to elements of the story in this book. But on their own, all of them miss some of the more subtle, but nevertheless profound, changes that tech is bringing. This book is not about the revolutionary things tech is enabling nor about the singular figures leading the industry. Instead, it is about how technology is shaping how we understand and experience the world, rewiring basic concepts such as friendship, conversation, and memory. To understand this, it is important not just to look at any given technology in terms of its starting point or its end point, but instead at its long-term cycle. Mobile technology, more than any technology before it, exists in a constant and quick cycle of refinements and user feedback, greatly accelerating not only the development of products, but also the ways those products affect us. Further, it is the nature of digital technology to accelerate these processes even more as our phones constantly intrude on many of our lives, with notifications and reminders pressing us to engage. This book examines digital technologies from a sharply different perspective than the valuable critical contributions that describe and explain digital technologies primarily through their economic, social, and political effects. Instead, we focus on the meanings and cognitive structures that we will show are profoundly altered through digital technologies. Digital technologies not only produce new tools and social structures, but they produce new meanings; they change how we see things. Like creative metaphors that make us see one thing through the lens of another, we argue that digital artifacts create and reconfigure meanings and cognitive structures through the use of metaphorical processes.

At the center of the change produced by digital technologies is inevitably a metaphor. As we shall see, metaphors in technology allow developers to take esoteric concepts and make them meaningful to people. Once they have made that connection in people's minds, e.g., that following someone on social media is a form of friendship or that texting someone is a kind of conversation, then the process of the technology changing how users think has surely begun. Thus, as text messaging has evolved over the years since its clumsy beginnings, so too, in parallel, has our notion of what it means to hold a conversation with someone. For many of today's teenagers, for example, a "conversation" is more likely to be held by text than it is face-to-face. And for those teenagers, every time their phone buzzes with a notification of a new message from a friend, that cognitive change is reinforced. Meanwhile, there is a parallel story unfolding, one of people without access to such technology, whose understanding of the world is increasingly different from those with access.

Since our main goal in this work is to examine how technology is changing how we think, we will examine the very concept of meaning, as well as the cognitive mechanisms involved in making it. Crucial to all of this is the role that metaphor plays in how we learn new ideas and how we communicate. The book is born of an interdisciplinary collaboration between a philosopher and a cognitive scientist, both of whom are steeped in technology. As such we will bring elements from both of these areas to bear on the ideas in this text. However, our goal is that any educated person with an interest in how technology is changing how we think should be able to understand the arguments in the book. Since the technologies that our book is primarily interested in come out of the mobile phone industry, many of our examples draw from that industry, including the phone itself. Telephones, and our relationships to them, are still evolving more than one hundred years after their invention.

There is a romantic popular notion that helps reveal the public's fixation on how new technologies and inventions are created. The image, probably dating back to at least Thomas Edison, dubbed "the wizard of Menlo Park," is of a solitary genius, wearing a white coat, hard at work in his lab who is suddenly struck with inspiration. Our hero may even mark the moment by shouting "Eureka!" In this romanticized—and sexist—view of innovation, new technologies are the product of inspiration, often springing fully formed into someone's consciousness. The symbol of such an event is a light bulb. Further, the greatness of this innovation is so self-evident to the public that it is an immediate sensation. The reality, of course, rarely matches this imaginary scene. Even the lightbulb, the iconic symbol of inspiration, has a long lineage predating Edison's work. Edison himself, and notably a team of researchers, tested more than three thousand designs before submitting a patent.[1] This imagery is a kind of metaphor itself and the source of a great deal of power in terms of its role in setting expectations. For example, in 1957 surveys showed that people stereotypically believed that a scientist "is a man who wears a white coat and works in a laboratory."[2] Such metaphors are still going strong today, constantly reinforced in various media, and apply to many of the areas covered in this book.[3] As we shall see, such metaphors are powerful because they can subconsciously change how

Figure 1. An iconic metaphor, with a twist, suggesting sudden inspiration. The lightbulb metaphor is so powerful that innumerable variations can easily be found on the internet. In turn, having the idea come from a woman of color is in opposition to the metaphorical idea of "male genius." (Figure by Kira Chown)

people behave, e.g., young girls may be less likely to pursue science because they do not fit this metaphorical image.

When a technology is created its developers have a choice in terms of the metaphor that they will use to describe it. As with the image of the lone scientist in a white coat, such metaphors frame the way that people will see the technology and influence how they react to it. Metaphors are not just about technology; they are the main and most important way that we understand technologies. A good metaphor can help make a technology a success, while a poor one can hinder its use and development.

In examining the relationship between metaphor and technology, we must first start with a working definition of what we mean by technology. As we will see later, in chapter six, carefully defining categories is a fraught endeavor, and so, despite some reservations about this particular definition, we employ a working definition:

Eric Schatzberg's description of technology as "a set of practices humans use to transform the material world, practices involved in creating and using material things."[4] We also find it productive to complement Schatzberg's broad definition with Luciano Floridi's practical intuition of technological artifacts, which relies on the notion of "in-betweenness."[5] A window is between you and the outside world and is thus a technology. A saw is between you and the wood you want to cut and so too is a technology, etc. There is a hierarchy to this, as well; a hammer is between you and the nail you want to drive into wood, and thus a technology can be between you and another technology. What we like about this particular notion of technology is that it captures the fact that technologies are mediators of experience. While Floridi's examples emphasize the practical aspect of technological mediators, we will analyze the cognitive and interpretative dimensions of these mediations. We should immediately note that popular notions of technology, and indeed this particular definition, focus on technologies that exist in the physical world, but many of the cases we describe in this book deal with software technologies that are only made manifest through other technologies, generally mobile phones.[6]

Since much of our book deals with mobile technology, and specifically technologies that are "enabled" by smartphones, we will start with the example of telephones, and we will weave metaphors relating to telephones of all kinds throughout the book. This concept of "enabling" is crucial to our discussion of technology and to understand how technologies come to be. The idea was brilliantly laid out in the PBS television show and companion book *Connections*.[7] Each episode of the show began with a historical event, generally hundreds of years prior, and subsequently traced a series of connections back to some product in the modern world. The links in the chains of connections can be seen as technologies in their own right and/or as enablers giving rise to the ultimate invention or product.

Thus, there is a predominantly technical process of developing artifacts and architectures that allows certain features of new technologies to be implemented. These processes often take place in public and private research institutions and focus on developing what we call "enablers." In the case of the original telephone, the coding of audio signals as electrical pulses and the infrastructure for transmission of such signals are among the basic technology enablers.

Even if they are fascinating in themselves, we won't focus on these basic technology enablers in this book. Our main interest will be in technologies that pass one key criteria: they should be immediately recognized by their users as creating new possibilities for them to perceive and/or act in the world. The act of converting audio to electric pulses, for example, is not immediately perceived by a user making a phone call in a way that they would understand. Thus, technologies depend not only on enablers, but also on meaning. Coding audio signals into electrical pulses is a technical achievement, one crucial to telephones, but it is essentially hidden from the people who use telephones. By contrast, a mobile phone is both an enabler, in that it allows for the development of new technologies that build on it, and a full-fledged technology in the sense we are talking about here, in that it mediates numerous human experiences. When enablers are not thought of in a way that assigns new meanings to our lives, they remain mere artifacts of scientific progress, not of practical interest to the greater public. The transition from basic scientific research to technology, at least in the sense that we mean here, comes from a semantic leap where new meaning is created through its usage in the world.

Just as technologies are not created from thin air, neither are the new meanings created by technologies. With technologies, creation relies on enablers, preexisting innovations with their own history. We argue that the new meanings brought by technologies are also not created out of whole cloth, but rather by what could be termed "cognitive enablers." These are the other meanings that already exist in the semantic universe of a given society. But what cognitive process is linked to this semantic innovation? How does it work, and how is it structured? Our supposition is that the semantic innovation process of new technologies can be understood in terms of the way we create metaphors. Metaphors act as a paradigm for semantic innovation in language. In this book we will explore how this process illuminates the way technological innovations happen in the digital world, and how it is often ignored or overlooked in comparison to the development technology of enablers.

When it comes to digital and computational technologies, there is already a substantial and rich base of enablers, the result of decades of development stemming from the computer revolution. This means that discussions of the development of new technologies in this sphere can focus squarely on semantic innovations. Enablers such as low-cost LCD panels may be worthy of discussion in their own right, but there is no way in which they are important with regard to, for example, the development of social networks.

Returning to our telephone example, it took an effort of creative imagination to bring two distinct conceptual entities together in the creation of a new concept, and therefore a new way of understanding the world. There is a large semantic gap between copper wires and electric shocks and the concept of human conversation, a gap that needs to be overcome for enablers to enter users' semantic universe and, therefore, to be understood and used.

The telephone was born as a metaphor for "conversation." Imagine a universe in which it wasn't, and instead prospective users were given a set of explanations about coding audio into electronic pulses, and how these pulses could be transmitted over wires, and later turned back into sounds. None of those things would have been meaningful or particularly interesting to the average person in the nineteenth century, who probably would have stopped paying attention. Instead, these details were abstracted away, and the telephone was presented as a way to hold a conversation with someone who was somewhere else. This is simple and easy to grasp, relying on concepts that we all share—that we know what conversations are and that conversations normally take place between people who are close together. As Katzenbach and Larsson put it, "imagining the future is always mobilizing the past."[8]

In his classic book *The Design of Everyday Things*, psychologist Don Norman noted that:

> Two of the most important characteristics of good design are discoverability and understanding. Discoverability: Is it possible to even figure out what actions are possible and where and how to perform them? Understanding: What does it all mean? How is the product supposed to be used? What do all the different controls and settings mean?[9]

In other words, for a product to be successful, it is of paramount importance that it make sense to its intended audience. In this book we are primarily concerned with technologies that are intended for the public at large, and thus need to be easily understood by people in general. This places a number of constraints on developers, generally requiring their products to be simple to grasp. This means that new products inevitably need to be cast in terms of things that everyone already understands; as we shall see, the primary vehicle for communicating about these new products is metaphor.

The set of cultural and fundamental knowledge that a person or group of people has before encountering a new technology is what we will call their "semantic horizon", roughly a combination of their experience and their knowledge.[10] This semantic universe can be implicitly and explicitly used by innovators, primarily through metaphor, to help give meaning to new technologies, ultimately easing the integration of the new products into the user's understanding of the world. The result is that new technologies that do this successfully are far more likely to be useful because they are far more likely to be used.

In conversation or when reading, a metaphor like "time is a beggar" is an "impertinent predication" that needs to be solved in order to make sense. One might wonder, for example, what features that time has in common with beggars, or vice versa. The metaphor only makes sense once we can work out the connections between the metaphor's tenor, in this case "time," and its vehicle, in this case "beggar."[11] As Norman alluded to, to be effective, technological metaphors cannot be quite so impertinent; they should be easily transparent to nearly everyone. For example, when we tell someone "I am going to talk to my mother on the phone," and the friend knows that our mother is eight thousand kilometers away, the friend's understanding relies on how using phones expands, modifies, and builds on the concept of "conversation."

As this example shows, technological metaphors fundamentally alter cognitive models and meaning. A new set of associations is created and incorporated into our cognitive models when a given technology becomes common in any social group. For example, the idea that conversations took place face-to-face was deeply associated with the concept of "conversation" when the telephone was developed. Once telephones became pervasive, it was no longer required that conversations take place in person, and thus the association between "in person" and "conversation" was necessarily diminished. A linguistic sign of this change is that, when we plan future conversations, we need to specify how they will take place, as in "let's talk in person," or "let's talk over the phone" and now "let's talk over Zoom." The changes that metaphors make on our mental models can have all kinds of consequences, both good and bad. Among them is the fact that metaphors often attribute agency when none exists. For example, studies of the stock market have shown that common stock metaphors like "the NASDAQ climbed higher" or "the NASDAQ dropped off a cliff" attribute agency in the first case and a lack of agency in the second to what otherwise might be seen as random fluctuations. This matters because the agency carries with it the expectation that the change will continue.[12] Thus, the choice of metaphors in such a case can affect how people behave after being exposed to them. Other changes to our mental models have to do with the associations the metaphor brings. For example, returning to our iconic light bulb, researchers found that subjects exposed to an illuminated light bulb performed better at solving problems requiring insight than those exposed to shaded bulbs.[13] The studies were carefully constructed to show that it was not the presence of light that made a difference, but an actual illuminated light bulb. This is taken as evidence that the metaphor primes the portion of the brain involved in insight, leading to improved task performance—because we associate light bulbs with creativity, seeing a light bulb temporarily makes us more creative!

Metaphors used to explain technology modify our understanding of the world, it is important to remember that technology also creates new social groups. People who participate in social networks and people who don't. People who have access to smartphones, high-speed internet, and the like, and people who don't. We will argue in this book that one aspect of technologies enabled by smartphones is that their impact on cognition is accelerated and heightened. As a result, the differences between groups that use or do not use a particular technology are growing quickly as their models of the world are diverging.

Adopting new technologies has important repercussions since the ability of potential users to understand the purpose of a new artifact depends on how it relates to their individual models of the world. This situation can be framed as one in which experts need to teach novices about their area of expertise. As Stephen Kaplan noted in an examination of such situations, in many cases where experts try to teach novices the result is frustration on both sides due to the different ways the two groups perceive the domain in question. Things that have become blindingly obvious to experts may be opaque to novices. As Kaplan puts it, "if one sees the world very differently from the way others see it and counts on others to see as we do, the consequences can be unfortunate."[14] This is central to this book in two ways. First, one of the challenges of creating a successful new technology is bridging this gap through the use of metaphor. Second, the groups that use specific technologies and those that do not will end up seeing the world very differently, just as in Kaplan's quote, and thus unfortunate consequences are almost inevitable.

Meanwhile, when a technology fails to make a connection to the public it is likely to fail no matter how useful it might be. An example of this is QR codes, a technology that frequently comes up when doing nearly any variation of an internet search for "biggest technology failures." The first problem that QR codes have is that there isn't an easy metaphor that explains what they are for. The best we can do is something like "QR codes are images that are hyperlinks." This is relatively accurate but does not give any sense of why they might be useful. QR codes also fail Norman's design rules. Looking at QR codes gives no sense whatsoever of what they do, nor does it help to understand their functionality. With the growing emphasis on privacy, this obscurity can also make users very uneasy about what will happen when they scan a code.[15] Indeed, QR codes are purposefully designed this way so that they can be integrated into camera apps without interfering with normal picture taking. QR codes continue to exist and have steadily grown in usage because they are in fact a useful technology that solves an important problem, especially for those who understand their technical side, but they have never found their way into the public's imagination.[16]

As people incorporate the use of a particular technology into their daily lives, the technology affects that group's semantic horizon. Technology that was born of a metaphor becomes part of the common lexicon and alters the meaning of the metaphorical "vehicle." Thus, the telephone that is born as a metaphor for "talking" (one can talk over the phone) changes the associations of "talking" so that the verb "talk" can also be used to mean a distant communication.

For users of a new technology, such as the telephone, the device spurs the creation of a new semantic entity, or category. This new entity in turn uses metaphor to leverage another set of prior experiences to help understand it, e.g., when the telephone was invented, its usefulness could be sold to people on the basis of their understanding of conversations and the limitations on how conversations had previously worked. Once they had a telephone, however, the device became a new semantic entity for them, one centered around the actual experience of using real telephones. This experience is crucial to our story because experience is a key driver for learning and altering our semantic horizons.

Interestingly enough, once the new semantic field for a new technology is established, the field itself can function as a basis for further metaphors. Thus, for example, the mobile phone metaphor is born from the telephone, further removing restrictions on how conversations take place. A new technology can therefore enter the semantic field as a metaphor for a different technology.

It is interesting to note how Albert Borgmann's idea of separation between means and ends of technology is pertinent to this example.[17] In our terms, "means" are what we are calling enablers. The mobile phone metaphor is solely about the ends of both technologies involved—landlines and mobile phones. The means have no obvious effect on the metaphorical understanding. Indeed, mobile phone enablers are substantially different from those for a landline telephone, and so "means" are virtually ignored with regard to the semantic field of the new metaphor. Again, for users such information is an unnecessary distraction, even if it is crucial for the developers of the technology.

Mobile phones are also indicative of how creative imagination can have a practical effect on new technological metaphors. A fictional precursor to mobile phones, the "communicator," was popularized through the *Star Trek* television series, itself a kind of fictional testing ground that showcased the usefulness of such a technology and thus prepared potential users and developers for why they might want such a device. Viewers could indirectly experience, through the television program, how such devices could be useful. In practical terms, the *Star Trek* communicator inspired some models of early commercial mobile phones. The possibility that the metaphor of a technology might be created even before the enablers are actually available is yet another indication of the decoupling of means and ends, allowing the metaphor to be meaningful and fully understandable without the means being fully realized or taken into account for the resolution of the metaphor. Science fiction is full of such examples, e.g., lightsabers in *Star Wars* are easily understood as "laser swords," even though the technology necessary to create such an item is currently well beyond our scientific understanding.

Donald A. Schön explores the example of a technology-related metaphor from the perspective of its engineering team. He develops the metaphor that "a paintbrush is a pump." He says:

> The researchers, who had begun by describing painting in a familiar way, entertained the description of a different, already-named process (pumping) as an alternative description of painting and that in their redescription of painting, both their perception of the phenomenon and the previous description of pumping were transformed.[18]

This new way of seeing a paintbrush as a pump led researchers to new questions such as how the density and structure of the fibers of the paintbrush should be organized to better produce the pumping effect that ended up creating a set of technological inventions.

Jahnke has also explored how the creation of meaning is related to the innovation process. He argues that it plays a dialectical role with problem solving:

The point is that all the problem solving occurred within a process of seeking an evolving meaning. Interestingly, this experience corresponds with research in science and technology studies indicating that science and technology development is not as rational as it may seem. Imagination, metaphor, experiences, and other "irrational" thinking are necessary to coming up with new scientific concepts and innovations.[19]

Returning to science fiction, Orson Scott Card gives an example of the power of meaning and problem solving in his novel *Ender's Game*. In the book a character describes how a faster-than-light communication device dubbed the "ansible" (after another science fiction device first described by Ursula K. Le Guin) was developed after encountering an alien species that could communicate that way. "We knew then that it was possible. To communicate faster than light. That was seventy years ago, and once we knew what could be done we did it."[20]

The conversation metaphor did not stop evolving with the creation of mobile phones. Interestingly, though, the next steps of the metaphorical development were in software, not hardware. In the early 1990s when the Short Message Service (SMS) became available to customers, who now know it better by its colloquial name "texting," the "conversation" metaphor moved from audio to text. We will examine this development in much greater detail in a subsequent chapter but mark its development here to note that computers have helped change the meaning of "technology" itself. While it is true that software like SMS currently requires physical enablers, there is no SMS device on a mobile phone, merely the software implementation of one. In this book we are mainly concerned with such digital technologies and metaphors. Thus, we might say that social networks are metaphors for relationships and that specific networks like Facebook or Tinder are themselves technologies. Given that previous technologies were physically embodied, this raises significant questions about the impact, if any, this has on the metaphors and the creation of semantic fields based on those metaphors and technologies. We will address these questions throughout the book.

Digital technologies are essentially related to metaphors, but digital metaphors are different from linguistic ones in important ways. Linguistic metaphors are passive, in the sense that the audience needs to choose to actively engage the world proposed by metaphor. Returning to the Shakespearean metaphor "time is a beggar," the audience is unlikely to understand the metaphor without cognitive effort and without further engaging Shakespeare's prose. Technological metaphors, on the other hand, are active (and often imposing) in the sense that they are realized in digital artifacts that are actively doing things, forcefully changing a user's meaning horizon. Technological creators cannot generally afford to require their potential audience to wonder how the metaphor works; normally the selling point is that the usefulness of the technology is obvious at first glance. Shakespeare, on the other hand, is beloved in part because the meaning of his works is not superficially obvious and requires some thought on the part of the audience.

In the same way that there are distances between vehicle and tenor in linguistic metaphors, there is also a distance between artifact (tenor) and the vehicle in technological metaphors. Earlier we saw with the *Star Trek* communicator that a metaphor might seep into popular discourse well before its technological enablers were ready. However, it appears that in some cases this is not true. Mainly this arises when the enablers are not sufficiently developed for the metaphor to be satisfactorily matched by users. Examples of this currently include 3-D technology and its offshoots such as "virtual reality," which have frequently been touted as "the next big thing" but have yet to live up to their hype. The enablers have never been good enough for the artifact to be recognized in fact from its similarity to the vehicle (in this case, the reality as we experience it in everyday life). Virtual reality may be virtual, but it is still severely lacking in the reality department.

In this book we examine digital metaphors, their development, their acceptance into popular culture, and their impact on a culture's meaning horizon. To do so we must first dig more deeply into what metaphors are and how they are used by—and consequently impact—our cognitive systems. This examination will reflect our own backgrounds. One author is a philosopher who also spent twenty years in the tech industry, working for companies including Motorola and Google. The other is a cognitive scientist whose background in technology includes being the head of a world champion team of soccer-playing autonomous robots. Thus, our examination of technology will be grounded in philosophy and cognitive science, which will be addressed in Part II.

Part I of the book consists of four chapters. In those four chapters, we build a framework based on philosophical and cognitive research. We contend that the power of what we are doing comes from the blending of these two points of view. It is not our intention to break significant new ground in either discipline, but instead to find common ground that can be used to build a framework for understanding technology, particularly its relationship to metaphor.

This part starts with the topic of metaphor since it is central to everything in the book. Far from a mere artifact of language, metaphor plays a crucial role in how we explain novel things, how we conceptualize the world, and how we act in the world. Our goal in this chapter will be to show just how much of meaning is wrapped up in metaphors. Once that is done, we turn to an example that is meant to illuminate the main points of our book. This example, drawn from our own lives, shows how intrinsic technology is to modern life, impacting how we communicate, how we relate to one another and even how we remember events.

The subsequent two chapters present the crux of our approach. In the first, we turn to the topic of digital metaphors and how they are different from ordinary metaphors. This is primarily due to the fact that they are made concrete by being implemented on smartphones, but also because of the pace of updates that the digital world has enabled. We show that metaphors are central to the creation and deployment of new technologies. Digital metaphors are crucial in explaining the usefulness of technologies to users who might not understand the technical details. These metaphors are then implemented and made material. This materiality, in conjunction with the widespread and constant use of smartphones, has the effect of supercharging the metaphor's impact on meaning.

Next, we show that in the world of mobile devices the technological enablers have created a situation in which there are very fast feedback loops between users and developers. This allows technologies and their underlying metaphors to shift very quickly. This process has the effect of redescribing users' models of the world. For example, people who regularly participate in text messaging exchanges will inevitably have a different model of what it means to have a conversation than people in the pre-mobile world did or, notably, people who simply do not have access to smartphones. And, of course, these changes in our models of the world have consequences in terms of what we believe and how we behave. As we shall see they can have legal consequences as well.

In Part II we connect our framework more closely with its philosophical and cognitive underpinnings. This part may, at times, verge on being too technical for some readers and such portions could be skipped, but it is essential for developing some of the theoretical concepts that we rely upon.

In the first chapter of Part II, we take on the concept of meaning using frameworks developed by philosophers. One of the primary difficulties in dealing with meaning is that a sentence can mean different things to different people based on their experiences. These experiential differences between people can mean that their models of the world are different. A simple sentence like "Rio de Janeiro is a city of contrasts." will have vastly different meanings to someone that has only visited Rio briefly, someone who has lived in Rio their whole life, and someone that has only read or watched news about Rio. This will be important with regard to technology because not everyone has the same access to technology. In trying to untangle this, philosophers have created systems for analyzing sentences to try and get at their underlying meaning. For example, among the important ideas in this work are discussions of what philosophers call referents and predications. We can roughly think of a referent as the thing in the world that a word is referring to, and predications as particular features or actions that a sentence is describing or emphasizing. In our simple sentence, the referent would be Rio, and the predication would be the description that it is full of contrasts. This may seem similar to diagramming sentences in grammar school, but it turns out to be a complex topic that is full of nuance thanks to context, whether that context is cultural, part of a longer conversation, or the internal mental models of the people in the discourse.

Meanings are not just language structures without practical effects; they are lenses through which we view and make our daily choices, from the most trivial ones, like what we eat for breakfast, to the career we intend to pursue. This discussion of meanings will form the conceptual background in our analysis of metaphors as the fundamental mechanisms that evolve meanings over time. Metaphors are not merely linguistic embellishments that make sentences prettier. As Lakoff and Johnson, and many others in the wake of their seminal work, have shown, metaphors are basic mechanisms for creating meanings.[21] When analyzing metaphors as semantic and cognitive phenomena that transform and create new meanings, their profound importance for the understanding of cultural and social processes, such as digital technologies, becomes manifest.

In Part II's second chapter we look at how knowledge is structured and acquired from the perspective of cognitive science, paying particular attention to how this perspective intersects with the ideas put forth in the first chapter. In a rough sense the referents in the first chapters are cognitive objects, or the categories that form the basis of a significant portion of cognitive research. The complexity of categories is reflected in years long internet debates about things such as what constitutes a sandwich. As we noted with Rio, any given person's categories will reflect their own experience of the world. The mechanism for translating that experience into cognitive structure is Hebb's rule, which describes how neural connections are strengthened in the brain. The main result of this rule is that people associate things together when they experience them closely in time. These associations not only form the basis of categories, but they also link the categories together into a kind of network, which in turn is itself the basis for the predications discussed in Part II's first chapter. Normally this process is fairly slow and statistical, but it can be sped up in at least two important ways. One is through intense emotional experiences. The second, our main area of interest, is through metaphor.

Part III of the book consists of several chapters that look at specific technologies from the perspective of our framework and how these technologies are mediating essential parts of our lives. First is a chapter on what may be the most successful technological innovation of the twenty-first century: the touch user interface. When Apple announced the first iPhone in 2007, a great deal of the conversation was not centered on its features but on what it didn't have—a keyboard. The success of the iPhone can be largely traced to what many, if not most, tech observers saw as its greatest weakness, and that success was predicated on a powerful metaphor, that "touching is selecting." The metaphor meant that there was no learning curve needed to use this new device; even toddlers are used to "selecting" things by touching them. Over the next decade the metaphor was so powerful that it became ubiquitous, forever changing how we interact with technology.

Next comes a chapter on a technology that has dramatically changed what it means to have a conversation with someone. SMS is an example of the hermeneutic cycle in action. The original metaphor, based on a combination of mail and telegrams, was poor, and early implementations were limited by the available technology at the time. Over time, however, as the technology quickly improved and with a growing user base that pushed the software in new directions, SMS became the most used app in the world and the metaphor shifted from mail to conversation. SMS and its successors have changed the nature of what it means to have a conversation, shedding the bonds of time and space restrictions. The ability to have conversations that don't require being in the same place at the same time has obvious advantages, and those advantages have driven the success of the technology, but they also bring drawbacks that must be reckoned with.

The third chapter of Part III looks at how technology has changed our perspective on friendship through social networks. Social networks have also changed how we interact with other people, and along the way are changing what it means to be a friend and how friends catch up with each other. Because mobile devices have the ability to remove traditional restrictions on time and place, they afford new ways to do old things. Instead of catching up with a friend by having dinner with them, we can instead go onto their social media and read their posts, often liking them and posting our own replies. Friendship, therefore, has become a much more lightweight concept. Meanwhile, companies like Facebook can directly manipulate what it means to be a friend in their choice of iconography, response options, etc. At the same time, these companies are competing for their users' attention, doing everything they can to keep users online and connected.

The fourth chapter of Part III examines how digital photos and their associated apps have changed our relationship with memories. Pictures have always been a kind of proxy for memory, but now that everyone has a camera in their pocket all of the time, the relationship has strengthened and changed. This is further exacerbated by apps. Some of these apps allow us to modify our pictures. Are we also modifying our memories of those events? Other apps use artificial intelligence to try and concoct memories for us.

Both Google and Apple's Photos apps even name these collections "Memories." In this chapter we examine this ongoing evolution and look at growing evidence that it has an impact on how we view the past.

In the final chapter of Part III, we take a closer look at the device that pulls all of this together, the mobile phone, and ask whether we even have the correct mental model of what the device is. Mobile phones have changed our world. They have removed traditional restrictions of time and place on a myriad of human activities and enabled an endless number of new technologies that were not previously so easily accessible or even possible. Further, an aspect of mobile phones that is critical to this book is that they allow for frequent, simple, software updates, thus meaning that the development cycle of mobile technologies is orders of magnitude faster than that of traditional technologies. Mobile phones do so many things that calling them "phones" at all seems almost laughable. To that end some have taken to referring to them as "pocket computers," but that comparison too is lacking. Since the release of the iPhone, which kicked off the smartphone era, mobile phones have had their own cycle marked by relentless incremental innovation on the developer side, and a user base that constantly finds new ways to use the technology that developers did not imagine. Not only has this changed our model of what a phone is, it has legal consequences as well. Even as we write this, Apple has become embroiled in legal battles over the definition of what this device is, and therefore what kinds of controls Apple can exert over them.

Finally, in Part IV, we devote four chapters to the impact of digital metaphors on the world around us. Sometimes digital metaphors fail. This can be due to failures of imagination or of keeping up with the fast pace of technological change; they can also be ethical failures. We will examine such failures in Part IV's first chapter. Meanwhile, given the impact of digital metaphors on the world, it is crucial to pay attention to who is creating them. This is the topic of Part IV's second chapter. Failures are not the only potential problem of digital metaphors. Digital metaphors may be even more problematic when they succeed. In Part IV's third chapter, we look at the consequences of success. Finally, in the fourth chapter of Part IV, we look forward, drawing together what we have learned to suggest a new approach to examine digital metaphors generally.

PART I

METAPHORS, TECHNOLOGY, AND MEANING

Is social media changing the very nature of friendship? What about text messages and how we communicate with one another? Are our vast collections of photos and videos changing how we remember our lives? If our goal is to answer these questions, and it is, then we must begin by examining how metaphors precipitate such changes. For users of a new technology the story inevitably begins with a metaphor. New products or apps must describe themselves in ways that are easily understood, and metaphor is how it happens in nearly every case. So we begin with metaphors and their importance in how we communicate and how we think. Once we have done that, we stop for a brief interlude, where we use an example to walk through some of the important premises of the book, making the case that the metaphors of everyday digital apps are fundamentally altering basic human activities. We then transition to digital metaphors and the special characteristics that they have, given that they are implemented in technology. Finally, we look at how this combination of a digital metaphor and a networked device has a transformative power the likes of which the world has never seen before.

CHAPTER ONE

METAPHORS

A METAPHORICAL WORLD

"Her voice faded away." "Put the whole episode behind you." "She sailed through her exams." "She poured out her problems." "We dug up some interesting facts." "You can use the mouse to add it to your cart." "How many likes did you get?" You would be forgiven if you hadn't immediately realized that all of these simple phrases are metaphorical;[1] they all ask us to see one thing as something else. According to these metaphors, the sound of a voice is something that can be seen depending on its volume, facts from the past are physical objects located behind you, examinations are a sea to be navigated, a digital pointing device is a rodent, pressing a button is a token of appreciation, and so on. Metaphors are so common in natural languages that some go so far as to claim that they are our most fundamental mechanisms to express meanings about the world and our experiences.[2] As we shall see, metaphors have a life cycle. When we encounter a metaphor for the first time it might require a moment's thought to determine what it means. But, as a metaphor becomes conventional through constant social use, it becomes transparent in the same way that wearing glasses modifies what we see even though we aren't actually aware of them in perception. They change how we see the world and, as we get accustomed to them, they become a natural part of our worldview.[3]

It is the "seeing as" quality of metaphors that makes them so integral to complex technology. If we had to truly understand every piece of technology before we could use it most of us would be very limited in what we could accomplish. Fortunately, we don't have to understand the minutia of how transistors or liquid crystal displays work in order to use a smartphone. We can see it as a way to hold conversations with our friends when they are not there or as a way to capture memories. Thus metaphors are important to our story in at least two ways. First, they provide a way for the developers of new technology to tell a story to the public about what the technology does and why it is useful. The second way is more subtle but at least as important. Those metaphors frame our relationship to the technology, and as we use the technology, in turn they alter how we see the world. The social media metaphor of friendship may become transparent to us, but it is still impacting our conception of what friendship is and how friends interact with one another. When a Google AI takes some of our photos and turns them into one of its Memories, it might seem like a cute trick, but interacting with those Memories impacts how we remember the events captured in the photos. In turn technologies typically work as amplifiers. A hammer amplifies our ability to hit things, a microphone amplifies our ability to speak loudly, etc. As we'll see this also works with learning, as one of the impacts of digital technologies is that they speed up the learning process and thus the impact of the metaphors involved.

Mark Coeckelbergh has highlighted the importance of language for studying technologies. He suggests that, while other approaches take essential steps in analyzing how technologies mediate our experiences in the world, they continue what Carl Mitcham has classified as an "engineering" approach to technologies in contrast to a "humanities" approach.[4] In particular Coeckelbergh argues that an important and all too common shortfall of various studies of technologies is not paying enough attention to the role of language. Exploring the diverse ways in which we relate to the world through language and technologies, Coeckelbergh concludes that "language is inseparable from technology" and that "technologies and languages shape how we think and speak about the world and shape what our 'world' 'is.'"[5]

This conclusion is an excellent starting point for our discussion, as it emphasizes the ways in which the intertwining of language and technology affect how we understand the world, evaluate our experiences, and prioritize our actions. In this book, we highlight the role of metaphors in this intricate relationship between us-language-technologies-world. Metaphors are privileged ways in which language and technologies blend. Metaphors can be used to create new meanings and are therefore an essential mechanism when developers want to introduce a new technology through language.

SEEING METAPHORS

We Speak through Metaphors

It speaks to the importance of metaphors that they have long been a focus of attention for philosophers and scholars of language going back to the ancient Greeks. Aristotle, for instance, recognized the peculiarity of the form of thought expressed by metaphors: "To metaphorize well implies an intuitive perception of the similarity in dissimilars."[6] Aristotle explored two fundamental characteristics of metaphors that are intrinsically linked to the essence of human language: persuasion and creativity.[7] In his course on *Rhetoric*, the Greek philosopher emphasized the use of metaphors for explanation and persuasion. He highlighted the emotional effects of metaphors, noting that solving the conceptual puzzle presented by a metaphor brings intellectual satisfaction as we learn something new about the world.[8]

Second, Aristotle, in *Poetics*, discussed how metaphors are ways of exploring and presenting new meanings. To show this he analyzed the creative process of tragedies that were considered the apex of artistic creation in the Greek polis of the time. To metaphorize well is to "see-as," to "perceive the similarity in dissimilars," remarked Aristotle.[9] Aristotle's analysis of the poetic use of metaphors invites us to think of their use as a creative process that builds approximations between two things that are not alike, and by doing so unveils, proposes, and communicates new meanings through language.[10]

Superficially, describing metaphors as "seeing-as" seems to be itself metaphorical. The situation is somewhat more complicated, however, as it appears that some metaphors, particularly spatial metaphors, are processed in part by the visual system.[11] At the very least metaphors in general seem to take advantage of, or at least mimic, the brain's ability to visually categorize, which often involves literally seeing one object as being the same as another. For example, when we see a brand-new cat, we see it as being the same, in this case meaning as a member of the same category, as other cats that we have seen before.

Unfortunately, after Aristotle the study of metaphors went through a long period of impoverishment. It was absorbed by the field of rhetoric, and thus for centuries metaphors were seen as mere figures of speech (tropes). This extended period, which began in early Middle Ages and extended into modernity, found metaphor studies focusing almost exclusively on formal aspects of how metaphors are linguistically constructed. Unfortunately, this ignored the creative dimensions of metaphor that have more relevant, exciting, and far-reaching consequences. Throughout this significantly long stretch of time, metaphors were considered mere word substitutions that could be used for aesthetic function, but that did not add or reveal new meanings. Some features of this reductionist view of metaphors persist to this day, such as when metaphors are listed as just another bullet point on a Pow-

erPoint slide full of figures of speech. This simplistic view of metaphors was based on variations on two interrelated premises: (1)
metaphors were linked with isolated words, and (2) the idea that
it would always be possible to replace any metaphor with other
words while maintaining the same meaning as the metaphorical
expression.

It was not until the mid-twentieth century that there was a
renaissance in metaphor studies, mainly through works at the
intersection of linguistic studies and philosophy. Important examples include the writings of I. A. Richards, Max Black, Monroe
Beardsley, and Paul Ricoeur.[12] These thinkers shifted the study of
metaphors by breaking with the two premises mentioned above.
First, they pointed out that metaphors are not exclusively linked
with isolated words but interrelate complex thoughts and contexts.[13] When someone says "data is oil," the metaphor relates not
only to the two specific words, but also to the set of thoughts and
implications related to data and oil, such as forms of production,
economic value, and environmental impacts. With this, these
thinkers shifted the focus from how metaphors affect the style of
what is expressed (rhetoric) to how they affect meanings (semantics). Furthermore, and most importantly, they also challenged the
second premise: other words with established lexical meanings
could easily replace metaphors without semantic loss. Thus, there
is something unique that is gained in terms of meaning by using
"oil" and not just saying that data is valuable. Both the source (oil)
and the target (data) of the metaphor operate in tandem to generate meaning. From the interaction between the cluster of significa-
tions evoked by data and oil, new meanings emerge to make sense
of reality and experiences.[14]

This semantic perspective is what most interests us in this
book, because it enables an analysis of metaphor as a rich cognitive process that can be central in the creation of new meanings.
Paul Ricoeur suggests that this ability of metaphor to be used for
semantic innovation is born of the exercise of an intense cognitive

activity that he calls, borrowing from Kantian epistemology, "creative imagination."[15] A metaphor can redescribe our experience in the world, opening a fissure in the predetermined structures of established linguistic lexicons. If you were to encounter a Shakespearian metaphor that Ricoeur was especially interested in, that "time is a beggar," you may initially be confused and think it doesn't make sense. How could time be a beggar? Time isn't even a person! But the opposite is the case. Shakespeare is (literally) making sense; through the text (or dialogue, in this case) he builds a new set of predications for us and redescribes our understanding of time. In Shakespeare's imagination, we must see time differently, as a person who takes things of value from us, puts them away, and moves on, sometimes without looking back. After the surprise involved in seeing things in this new way, a realization may come: I had never thought about time in this way! We will revisit this metaphor, including the original text, in chapter six.

Commenting on Robert Frost's passion for metaphors and their critical role on our thinking, Judith Oster said that "great metaphors enlarge our thinking and our imaginations as we 'play' with their possibilities, but also test their limits."[16] When we think about the potentialities opened up by the metaphor "time is a beggar," we play with new possibilities for understanding our experience of the time that passes, and, as a precious by-product of this interpretive process, we reflect on our own relationship to time by changing the way we see the world and think about our relationship with it. Through this metaphorical game, we can think differently, explore new facets of reality, imagine alternatives to insufficient meanings, and make sense of new human creations.

In this book, we will explore this innovative potential of metaphors. They allow us to speak something new about an experience that was, until then, ineffable. Thus, our focus is not on the details of linguistic classification that make subtle differentiations between distinct figures of speech. Instead we want to explore metaphors as a paradigm of semantic innovation. Metaphors work their magic through a fundamental trick, a form of cognitive and linguistic illusion. They ask us to see one thing as another. This game of "see-as" is fundamental for us to understand how metaphors say something new about the world.

Such novelty brought about by metaphors should be understood in two ways, one relative and the other absolute. A metaphor may be expressing something new in relation to a particular listener or reader. We are saying something new to someone who may have never considered a certain aspect of reality or didn't know anything about it. In this case, we use metaphors pedagogically. We say something new to that person, but it has already been thought of and is well known by many other people. It has a new meaning in relation to that person. For example, if a child who has a cat as a pet sees a raccoon for the first time, that experience is new to them. To provide meaning to what they see, we can use a metaphor: "That's a masked cat." We are asking the child to see that animal as a cat, which she already knows well, but a cat "in a mask." The "see-as" is used to teach something new in that child's experience, but which is already well known by others. There is already a word in the dictionary for the masked cat, and, little by little, the child will replace the metaphor with the lexical form "raccoon." Still within a relative perspective, metaphors may aim to persuade through comparisons that arouse new perspectives and emotional content. For example, political speeches are often interspersed with metaphors. Without going into a critique of specific content, it can be said that metaphors in this context maintain their pedagogical character.

However, there is another use of metaphor that we also want to consider: metaphors that refer to completely new meanings that are not in any dictionary yet, something new in the world that is still orphaned in language. In this sense, metaphors can bring a new experience to language. This is what we are calling semantic innovation; metaphors can create a new meaning for something that could not otherwise easily be expressed. A brief inspection of the dictionary reveals the large number of terms that started as innovative metaphors and, over time, became just other words in the dictionary. Texts like this book have bodies, headers, and footers, for example, terms that started metaphorically but gained their own place as words.

Metaphors are, therefore, fundamental to the capacity and flexibility of natural languages to account for new natural, social, and technological phenomena. We certainly speak in metaphors. But the importance of metaphors is not limited to language, although that would be enough to highlight their incredible relevance to digital technologies. Starting in the 1980s, with the expansion of the cognitive sciences, metaphors also became intricately related to how we organize our thinking.

We Think through Metaphors

The renewal of interest in metaphor within more comprehensive semantic and interpretative perspectives has brought many benefits. We are particularly interested in studies of metaphor done in cognitive science.

George Lakoff and Mark Johnson, among others, have been exploring the crucial role metaphors play in the way we structure our conceptual system. In their highly influential book *Metaphors We Live By*, they suggested that "primarily on the basis of linguistic evidence, we have found that most of our ordinary conceptual system is metaphorical in nature."[17] Since the publication of their work, metaphors have been highly studied in many corners of cognitive science, but our interests are most closely tied to the linkage that Lakoff and Johnson make between the human conceptual system and metaphor. As Tendahl and Gibbs put it, the traditional view had been "that metaphorical meaning is created de novo," but "in the past 25 years, various linguists, philosophers, and psychologists have embraced the alternative possibility that metaphor is fundamental to language, thought, and experience."[18]

The basis of the human cognitive system is categorization. Categories allow us to make predictive models of our world. When we encounter a new animal, for example, recognizing that it is a cat allows us to make a set of predictions that are likely to be fairly accurate even if we have never seen that individual cat before. We have already noted that metaphors allow us to see one thing as something else. This is the very same thing that we do in perception when we see the new cat as being just like other cats that we have previously encountered. A difference between the two cases is that with metaphors the "seeing" may itself be metaphorical, as we previously discussed. We must also note that with categorization, particularly for children, sometimes the new animal really is new. A child might be told, for instance, that what they are seeing is not a cat but a racoon. This is another challenge for some kinds of metaphors; we are told two things are the same when our cognitive system would naturally keep them separate, but as the example with cats and racoons shows, this is something people experience as part of learning categories.

Although there is no widely agreed upon model of categorization in cognitive, there are general aspects that are fairly well understood. When we see an object, we break it down into a collection of features. Depending on which features we see and their spatial relationship to each other we determine what the object is. A popular descriptive model of this, owing to the work of Eleanor Rosch, is called prototype theory.[19] What makes metaphors tricky from a cognitive perspective is that the features are implicit. When we are told "man is a wolf" we do not see a man or a wolf so that we might do a visual comparison. Instead, we have to use other means to determine what the relevant features are. In many ways this openness is where the power of metaphor comes from. When we see a new animal, our eyes constrain how we might classify it, narrowing the possibilities to what we directly perceive. With a metaphor, it is up to us to determine the relevant features. The question of how we resolve this process is the focus of much of the research on metaphor in cognitive science.[20] For a review, see Holyoak and

Stamenkovic, who especially highlight Walter Kintsch's model.[21]

In Kintsch's model, meaning is represented by a vector in a high-dimensional feature space. For Kintsch, resolving a metaphor involves merging the source and the target of the metaphor. There are several aspects of this that are important. First, for any individual the meaning vectors are going to be based upon personal experience, and thus the results of the process will vary from one person to the next and even more so from one culture to another. Second, it highlights the fact that metaphor is much more than simple redescription. Indeed, it can be creative, building new cognitive structures. Finally, the metaphor may create an entirely new cognitive unit. In Kintsch's model the merging of the two vectors creates a new one. We will dig more deeply into the shadings of this in chapter six, but the idea that metaphors can create new cognitive units, which in turn can serve as the sources for other metaphors, will be important to our story.

As with the example of a new cat, a metaphor provides a kind of template that can be useful for making general predictions. When Romeo exclaims that "Juliet is the sun!" he is telling us things about her that might guide our expectations if we were to meet her. We might expect her, for example, to be warm and bright and for any encounter with her to be pleasurable. These expectations give us a framework to work from when we first encounter her. But as with seeing the new cat, this framework cannot possibly be a perfect match for the real thing; this is even more true with metaphors because the source and target categories are often so different.

Despite this lack of fidelity, metaphors are cognitively critical because they allow us to provide a compact description to someone else in terms that person already understands. Thus, tech metaphors, such as friendship in social media, give us a set of features and expectations for something new even if we do not initially understand the technicalities involved. The importance of this was demonstrated during the coronavirus pandemic that began in 2020. One of the brand-new things to happen was the deployment of mRNA vaccines on a wide scale. This presented a messaging problem for a number of reasons. First, from the public's point of view it was a brand-new technology. Second, it did not work like previous technology. To make this trickier, even the previous technology is not well understood by much of the public. Third, and most importantly, there was a great deal of fear in the public about vaccinations, especially given the speed at which they were developed, going beyond the anti-vax sentiment that always exists. All of this led scientists and news organizations to work hard to figure out how to communicate to the public how this new type of vaccine works and why it is safe. One health news site rounded up what it considered the best five explanations from various sources, all stemming from experts in epidemiology.[22] All of the explanations used metaphors.

The first was that the vaccines were like Snapchat messages that expire. Among the notable things about this metaphor is that it assumes that the reader knows how Snapchat works. The second was that the vaccine was like an email sent to your immune system that shows it how the virus works and then expires like a Snapchat message. Aside from complicating the first metaphor, the second adds the metaphorical idea of showing the immune system the virus, or at least it makes it more explicit. The third metaphor was that the vaccine is like a recipe that contains instructions on what to make. The fourth likened the vaccine to a musical score, but one that only contains instructions to play a part of the song. The final metaphor is that the vaccine is like a system that only plays the catchiest part of a song, just enough so that you will recognize the song when you hear the whole thing. These last three metaphors emphasize the idea that mRNA vaccines, unlike traditional vaccines, do not actually contain the virus, but rather a code for a particular part of the virus, in this case the so-called "spike protein." From these metaphors, we can glean that mRNA vaccines do not contain the virus, but some sort of representations of it and, further, that even this copy is only in the immune system temporarily.

These metaphors were created to give the public a framework with which to understand the new technology. The idea is to take things the public knows about and use them to explain things that it doesn't. In turn, reasoning about the new vaccines will be greatly impacted by the metaphor that a person is exposed to. If you are told that the vaccines are like Snapchat messages, then you will apply what you know about Snapchat, if anything, to the new technology. As Lakoff puts it, when given a metaphor, we use the source domain (e.g., Snapchat) to reason about the target domain (mRNA).[23] All of the mRNA metaphors were chosen to emphasize a few salient aspects of the vaccines, namely that they do not actually contain the virus and that they are not retained in the body for a long period of time.

Another notable aspect of several of these metaphors is that they rely on tech metaphors themselves. Email started as a metaphor where we were asked to see a collection of activities on our computer as being like sending and receiving letters through the postal service. In the 1980s when the technology was new, the email metaphor would have helped set expectations for how it worked. Knowing about letters would have let you know that email probably involved communicating with people at a distance. Decades later, now that the metaphor has "died" and many people have more experience with email than the postal service, email can be used as a source for new metaphors such as that for the vaccines.

When experts such as immunologists need to communicate with novices, the challenges are substantial. Experts are immersed in their domain and its terminology and are used to thinking about them in a deep way. Novices, on the other hand, may not even have the most basic grasp of terminology, let alone any kind of understanding of how things work in the domain. This means that to effectively communicate with novices an expert must forgo talking about the very things that make them experts, in favor of things that the novice understands. This is no easy task, as reflected by the fact that none of the mRNA metaphors are perfect. As Lynne Cameron notes, making the technical language of specialists accessible to non-experts is one of the primary roles of metaphor.[24] We also note that the flaws in the vaccine metaphors, and the failure of the public to understand how they work, have had significant real-world consequences.

As Blavin and Cohen have suggested, metaphors are also applied to legal thinking: "Courts and commentators employ metaphors as heuristics to generate hypotheses about the application of law to novel, unexplored domains."[25] Metaphors are not only used as a way to communicate legal decisions or to argue in tribunal disputes, but more profoundly, they frame the thought process of unexplored domains. As Blavin and Cohen have also noted, technologies typically demand innovative ways to make sense of existing legislation, applying metaphors in the process of analogical reasoning "when courts encounter new technologies not yet anticipated by the law, their reliance on analogical reasoning plays a profoundly important role in the application of proper legal rules."[26] Again, the choice and quality of metaphors has a real impact on the world.

Nevertheless, there is more to meaning than just a list of features and cognitive processes. There is also affect and emotion.

We Feel through Metaphors

The gifts metaphors bear are not only concepts and ideas from a cognitive and linguistic point of view, but also expressions of affections, feelings, and value judgments.[27] They communicate and shape emotional content. Lakoff, for example, explores the metaphor "my job is a jail" and the frustration and other negative emotions evoked by such metaphorical choices.[28] Choosing another metaphor, such as "my job is a gem," would express completely different emotional content, engaging the audience not only cognitively but also affectively in ways different from the original.[29]

Lynne Cameron, an applied linguist who studies empathy in

dialogue and interaction, highlights the emotional aspects communicated by metaphors. One of Cameron's most critical findings is that metaphors are used to modulate the emotional content of speech. Thus, a CEO can use sports metaphors to motivate their employees. Teachers can choose metaphors with positive connotations that emphasize the potential for improvement when giving positive feedback to their students. And people with opposing ideological viewpoints can choose positive metaphors or more combative ones depending on whether they are looking for ways to reconcile or want to foster further discord. Cameron suggests, for example, that a systematic metaphor commonly used in reconciliation talks is that "reconciliation is connection." From it, a bundle of other associated metaphors is produced, such as "Let's build new bridges between our perspectives," and "Our communities spend too much time isolated in themselves."[30]

Metaphors are also widely used to modulate emotion and affect surrounding new digital technologies. For example, at the end of a route, Google Maps offers a choice of five emojis with faces expressing different levels of satisfaction, each representing the perceived quality of the routing algorithm. The icons with facial expressions are vehicles, and the user's experience with Google's route suggestions is the tenor of the metaphor. In this case, the metaphor itself involves expressions of emotions, but the relationship between metaphors and emotions in digital technologies is even more subtle and pervasive.

Consider the various metaphors proposed by social networks to capture different aspects of communication—messages, posts, conversations, chats, reels, and so forth. Each of them carries not only linguistic and cognitive associations but also affective and emotional content. For instance, almost all social networks, imitating a successful Snapchat feature, have created a way to share stories that contains some combination of texts, images, and short videos. The metaphor "story" was intentionally chosen to convey a casual and fun tone. Snapchat also highlighted the transience of this type of message, limiting sharing to a short amount of time, suggesting that users need not worry about long-term implications.

The recommender systems at the heart of almost every social networking and streaming media application provide further pervasive examples of emotional manipulation. The media feed is to be seen as your personal recommendation system, something like when you are looking for the next movie to watch and ask your colleagues about recent films they have enjoyed. This is even extended into the individual media items. In *Irresistible*, Adam Alter describes numerous techniques used to increase users' screen time.[31] One of them explores the cliffhanger media phenomenon. At the end of an episode, the plot develops into an expectation about what will happen in the next episode, which triggers an anxiety process that sparks a desire for the next episode so that it can be resolved. As Alter reports, in 2012, Netflix implemented post-play functionality that automatically plays the next episode in a series at the conclusion of the current episode.[32] While it may seem subtle, this change is critical, because research in this area shows a psychological trend toward inertia, and the post-play implementation shifts the user decision from needing to actively choose to play the next episode to needing to decide to stop it.

We Act through Metaphors

The metaphor "osteoporosis of the seas" asks us to see the effects of ocean acidification on coral in terms of how osteoporosis weakens bones in humans.[33] It redescribes a complex and relevant phenomenon in order to change the way people think (and hopefully act) in the world. It prompts predications of fragility, care, and support that we associate with human osteoporosis. Thus, the metaphor invites us to see coral in the same way we might see loved ones with bone problems: that they require protection from the underlying causes, in this case increased levels of CO_2.

Wessel Reijers and Mark Coeckelbergh proposed the concept of narrative technologies, which points to the fact that digital technologies create narrative frames that shape our experiences.[34] So, for example, the ignition process of a car is a narrative framing that guides part of drivers' actions every time they decide to use such technologies. From the moment the driver enters the narrative proposed by the technology, their actions are temporally coordinated. "The driver enters the car, puts its seat in the right position, adjusts the mirrors, starts the engine, is given visual feedback about the amount of gas in the tank, drives away from the parking spot."[35] The car and the driver become characters in the temporal organization proposed by technology.

Metaphors play a crucial role in understanding how narrative technologies shape our actions. The metaphors of social networks, for example, change how we interact with other people. To wish someone a happy birthday, we write a comment on our friend's timeline. At the end of a Zoom presentation, we express appreciation by selecting a clap icon. Stars have become a common metaphor for quality of service contracted through carpool apps. When selecting a number of stars, we act through the metaphors implemented in digital technology. Meanwhile, this action may have substantial implications for the drivers who depend on good evaluations to receive new customers and keep their jobs.[36] In digital technologies, the new materiality of metaphors makes them affect not only what we say and how we think, but also how we act and relate to others and the world.

Another essential form of intersection between metaphors and action concerns regulations and public policies. The implications of digital technologies for the grand scheme of social and political things are increasingly apparent. The dissemination of information on social networks has been testing democracies; social movements are created through networks of contacts and overflow to physical presence, such as the Arab Spring demonstrations; critical social resources as remote education capabilities are completely tied to technological infrastructures. The realization of the social implication of digital technologies has led to a surge in public and legislative debates that are often organized through metaphorical themes. For instance, in the debate over the right of internet providers to use the type and source of digital content to control their services—typically called "net neutrality"—two conflicting metaphors are commonly used: "the internet is a utility" and "the internet is a road." Choosing one metaphor or another has far-reaching consequences for framing the problem and, therefore, for regulatory and legal actions that have emerged in public debate.[37]

NEXT: TECH METAPHORS

In an article published in 2000, before the massive penetration of smartphones, Marakas, Johnson, and Palmer highlighted an intriguing aspect of the relationship between digital technologies and metaphors.[38] They explored the use of anthropomorphizing metaphors to describe the interaction of humans with computers, such as "it reads," "it writes," "it is friendly," and "catches and transmits viruses." The authors pointed to the use of metaphors in which humans are the vehicles and technologies the tenors, thus creating conditions to simplify the understanding of the relationship with computers, but, on the other hand, potentially fostering false or inappropriate attributions towards computing technologies. Through metaphors, computers and digital things were integrated into our daily lives as helpers, secretaries, coaches, analysts, accountants, and reviewers. But the use of metaphors in mediating our relationships with digital technologies goes much further. Metaphors are not just *about* digital technologies, they are *within* them and are essential for these technologies to be integrated into our lives. Next, we present an example that lays out the basic arguments of this book.

SPEAKING OF METAPHORS[39]

Metaphors are a fundamental means of communication. To understand a metaphor requires us to see one thing as being something else.

Metaphors take advantage of our ability to see multiple different things as belonging to the same category. In order to do this, we need to be able to determine the critical features that the metaphor is highlighting. Sometimes this results in a new instance of a category or an altogether new category.

Metaphors frame our expectations for dealing with the metaphor's target. Thus, metaphors can be used to shape behavior.

Metaphors are a primary tool by which experts can communicate with novices.

CHAPTER TWO

TECHNOLOGIES AND MEANING

In this chapter we outline the main objective of the book, which is to examine the impact of digital technologies on meaning. We frame this examination in terms of an extended example. In the example, which comes from our own lives, we see how smartphones have changed the way in which we communicate, how we relate to each other, and even how we remember things. In short, they are changing who we are.

A CAR RIDE HOME WITH THE KIDS

One of the authors recently had the opportunity to give a ride to a group of six or seven teenagers returning from a delightful afternoon of paintball. Thinking about his own past, he imagined that he would need ear plugs on the trip—a trip that would take around forty minutes, not counting delivery stops and cordial greetings from parents waiting for their athletes to arrive home at the end of an adventure that had been planned for weeks. He expected that his passengers would be excited to brag about their deeds on the playing fields and that the funny episodes and exploits would turn the car into an echo chamber like a busy market on a Saturday morning. However, those expectations were completely misplaced. On one hand, he was relieved by the tranquility of the car. Had it not been for a few glances in the rearview mirror to ensure that everyone was still breathing, he would have forgotten about his passengers given the sepulchral silence that ruled throughout the journey. This initial relief gave way to a mixture of concern and curiosity. Had something sinister happened in the game that made the atmosphere among teammates intolerable? Did his presence loom so large that it prevented spontaneous conversation? Were the combatants so tired that they couldn't even find the energy to talk to their friends?

All of these worries were resolved in an unexpected way. He asked one of his sons about the reasons for the silence, "Why didn't you talk about the game on the return trip?" The young athlete's answer was accompanied by the kind of look that only a child can give a parent. "What do you mean, we didn't talk?" his son replied, not understanding what he clearly felt to be a ridiculous question. "I didn't even see the trip go by; we were talking constantly the whole time. Do you want to see how animated our Instagram chat was?"

THE EVOLVING MEANING OF "TALKING"

Talking, chatting, having a conversation all mean something completely different for the boys than they have for their father throughout his lifetime. For the boys, talking can mean sending typed texts to another digital address. Though it wasn't true in this instance, "chatting" doesn't even need to be synchronous. The author's young passengers were not noisy, as was expected by someone whose notion of talking is based on different life experiences. For the author, sound waves are a central feature of what he thinks of as "talking." Instead, the boys were typing furiously; they were sharing their stories, bragging, joking—in other words, the content was the same, but the format was new. Social networks, in conjunction with text messaging systems and mobile phones, have created a metaphor that quickly took over the main meaning of "conversation" for the boys, who are part of a generation that grew up always having such tools available. A side effect of this change, particularly because it has been so rapid, is that different groups, such as the boys and their father, may have very different models of what a conversation is depending on their level of engagement with technology (Figure 2). Thus, a simple sentence like "I was talking with my friends" has become ambiguous in its meaning, although, depending on one's life experience, it may not seem ambiguous at all.

It may seem like a change in the meaning of a word is not important. Language changes constantly, after all. However, we are not talking about a simple case of language drift or subtlety in how a word is used. In this case, we are talking about how language is reflective of, and has an impact on, meaning. It isn't just that we have new shades of what it means to have a conversation; it is that we literally are changing how we converse and that those changes are important for many different reasons. As we shall see, a conversation over text is very different from one held face-to-face. In some ways it may be better, but in other ways it may be worse. Since conversation is such a fundamental human activity, it is well worth examining those differences and thinking about their ultimate impact and reach.

One negative consequence of this divergence in meaning is that it can only accentuate differences between groups with access to such technology and those that do not. Let's return to our discussion of the author's car ride to see how. Imagine, for example, one more passenger in the back, one who did not have a smartphone. Depending on the rest of the group this passenger might have been left out of the conversation completely, or some passengers might have resented how the conversation played out. A conversation where Dad can hear is likely to be far different than one where he cannot. Of course, in this case someone was excluded—the author!

Figure 2. The different experiences of young people have changed what the idea of having a conversation means compared to older adults. Thus, a simple sentence like "we were having a conversation about school" will have very different meanings to the two groups. For example, an adult might wrongly infer that the conversation took place face-to-face and synchronously. (Figure by Kira Chown)

With the rise of digital technologies such scenes play out over and over in many different contexts with many different groups.

Various psychologists and sociologists of technology have identified and examined the personal and collective consequences of the kind of semantic phenomena we describe in this book. The questions and interpretations of the impact of this new semantic universe on an intrinsically human experience like conversation are diverse. Some researchers[1] warn that text messaging creates barriers to the integral development of empathy; others suggest that this phenomenon merely creates a new form of mediation and brings the benefit of expanding contact with interlocutors from other cultures.[2] Still other researchers, like Margaret Morris, look at ways in which this new way of "talking" can become a nudge for other authentic personal and interpersonal experiences.[3] Irrespective of analytical perspective, the consensus is that the impertinent predication "chatting is texting" successfully reshaped users' horizons of meaning through instant messaging and social networking applications.

The example shows the power of digital technologies, a power that goes far beyond mere tools used to enhance productivity or automate an existing process. Digital technologies are changing how we see the world, value our experiences, establish interpersonal relationships, and live in society. We argue that the hermeneutic-cognitive approach that we present in the next few chapters is uniquely suited to examine the ways in which digital technologies transform our horizons of meaning through experience and, as a side effect, create new cultural groups—those who have incorporated these changes and those who are left behind. In turn, the semantic differences between these groups will make communications between them increasingly difficult. At the epicenter of the societal transformation brought about by new technologies are digital metaphors.

THE EVOLVING MEANING OF FRIENDSHIP

The changing way that people communicate today is not the only way that our relationships with each other are evolving. For some of the kids in the car, a normal next step upon returning home would be to post details of the day on social media. Thus, instead of waiting until the next school day to tell their other friends about what happened, they can tell their online friends about it immediately. From the early days of social networks such as Myspace and Friendster, social media apps have used the metaphor that connections within the apps are friendships. It is common, for example, to hear people refer to someone else as a "Facebook friend." This metaphor is reinforced by the fact that our "real" friends are likely to be among the first to connect with us on those platforms. In our example these online friends and followers are likely to be a much larger group than would normally be filled in face-to-face.

For the kids posting stories and pictures of their adventures, there is a kind of pressure to present them in the best light, hitting the highlights and leaving out more mundane details. It is natural for anyone to want to put their best face forward, and social media apps make that easier than it might be in face-to-face conversation. Aside from wanting to share with one's friend network, there is also the desire to gain "likes" and thus social capital. Thus, friendship, in social networks' terms, can be performative, and the quality of performance can be directly measured in a variety of ways. Meanwhile, a kind of counter pressure happens when their friends see those posts. The way that one supports friends on social media is to hit that like button and possibly to post a positive reply. In the age of social media, face-to-face interactions are as likely to be about what is happening in that online world as anything else.

As was the case with conversation, we see that increasingly the basic activities of a relationship are mediated by technology. Friendships are managed on Facebook and Instagram, romance on Tinder, professional relationships on LinkedIn, etc. We will examine the impact of social media on relationships in more detail in chapter nine.

THE CHANGING SHAPE OF MEMORY

One effect of an exciting day like the one the boys had is that memories are made, and the nature of these too is shifting and evolving due to the effects of technology. And with memories, once again, we can see the power of metaphor in conjunction with technology. Our memories are stories about our lives. In 2013 Snapchat seized upon this idea and started using stories as a metaphor for a new feature that collected groups of pictures together. By posting a collection of pictures as a story, you implicitly said that the pictures tell the story of an activity in which you engaged. In turn, Instagram essentially copied the metaphor into its own app. Later, Apple and Google repackaged the idea as a more direct metaphor called "memories" in their photo management apps. In this case Apple and Google automatically select and group pictures for users, with the idea that doing so will help users to engage with their pictures more often. In social media, the metaphor is that the way you tell your story is through a series of photos, while for Apple and Google it is that these collections *are* your memories. Both of these rely on the fact that we all carry digital cameras with us everywhere we go. Our smartphones allow us to easily document our lives as never before. In turn we can choose, edit, delete, crop, filter, and do all manner of things to photos to enhance those memories.

A common, and mistaken, metaphor for our memories is that they work like watching a film of your life. In reality memory is much more constructive. We remember some details and fill in others to make a seemingly coherent whole. Further, the mere act of remembering, and especially of sharing, memories with others can powerfully affect our memories.[4] We'll see why this is the case in chapter six. It wouldn't surprise anyone to learn that looking at photos and videos helps reinforce our memories of things that we have captured, but it is also important to recognize that they too are altering our memories to better fit digital documentation. And since that documentation can be altered and edited, we are essentially willfully altering our own memories, mediating them with digital technology.

NEXT: FOUNDATIONS

This book consists of several parts. The first part of the book is about metaphor and especially about how metaphors are used in technology and the special properties that tech metaphors gain through their instantiation in technology. This part of the book will develop our major themes.

SPEAKING OF TECHNOLOGIES AND MEANING

Technology is changing how we communicate, relate to each other, and remember our lives, among other things, at a very rapid pace. It is the nature of technology that these changes reach some groups well before others. Understanding how these changes are happening and their consequences is central to this book.

CHAPTER THREE

DIGITAL METAPHORS

If you are reading this book, you almost certainly have a smartphone, and you likely have it with you nearly all of the time. The capabilities of smartphones grow in power and sophistication constantly. This combination means that we use them to do more things, and in turn we rely on them in more and more aspects of our lives. Despite that, most of us cannot begin to comprehend how they actually work. Is it the actual hardware or the apps that make smartphones so useful? On one hand, the hardware continually evolves, gaining complexity even as it shrinks. On the other, software too has become more complex as it is given more tasks.

Yet, paradoxically, most of us feel like we understand our devices quite well. You may not know how to turn a soundwave into an electronic signal or how information is moved around a circuit board, but you might feel comfortable in the feeling that such details are unimportant to you personally, and even if you do not many other people do. After all, we understand our devices by what they do for us. From this perspective a smartphone is a device that lets us hold conversations at a distance, update our friends—wherever they may be—on what is happening in our lives, and capture memories with its camera. Our understanding, such as it is, of these activities relies not on technical know-how, but on metaphors. Phone calls, texting, and email are all ways of having conversations. Social media allows us to connect and catch up with our friends. Cameras capture digital memories. All of these are based on things that we are familiar with and are simply updated, and potentially improved, in their digital incarnations. Digital conversations, for example, are not constrained by time or space, nor is the ability to catch up with friends. Meanwhile, digital memories are far more durable than real ones.

As we saw in the previous chapter, metaphors ask us to see one thing as something else. When Romeo says "Juliet is the sun!" it is an elegant way to tell us how he feels, but we also understand that he does not literally mean it. It would be easy then to say that, when someone calls texting conversation, they do not literally mean it either. That instead, it is just an easy way to highlight some features of texting that are in common with conversations. As we have already seen, with metaphors generally, and especially with digital metaphors, there is much more to the story. Metaphors have the power to shape how we think and how we learn. They also have the power to change our understanding of things with which we are already familiar. Since digital technology has become so pervasive in our lives, it has become imperative to examine just how digital metaphors are impacting us.

THE ROLE OF DIGITAL METAPHORS

As our example of the car ride home with kids showed, digital technologies can change how we think. They can create new categories and alter the meaning of existing ones. But before they can do this people must choose to use them based upon some sort of description of what they are. And, since those same users exist in a world where there are many choices, new technologies must provide users with answers to some basic questions. Two of the questions are related: What is it, and why should I want it? There is a third question that is also important when it comes to understanding something, but which we argue is generally overlooked with digital technology: How does it work? The answers to these questions need to be simple. Otherwise, potential users could just choose to move on to something else. After all, why waste time being confused when there are so many other options?

At the same time, explaining a new technology is difficult since it often involves technically complex issues. A developer's main goal is to get users interested, so explaining "how" is far less important than "what" or "why." This is good news for developers since "how" is likely the most difficult question to answer in a way that ordinary people will be able to easily understand. Thus, most developers make a choice, abstracting their product to a metaphor that ideally simultaneously tells a user what the product is and why it is useful. Thus, a mobile phone is a telephone that can be used to make phone calls from anywhere, a post on a social network is an easy way to let your friends know what you are doing, etc. As with any abstraction, many things about the product are left out in these metaphors. For users this is generally fine, as most are uninterested in, and/or do not have the background required for, understanding the technical aspects that are required to understand questions of how. For now, we will simply note that there are consequences to this abstraction, e.g., understanding how an app transmits information can have consequences for the safety of your data. We will return to those consequences later in the book. For now, we focus on what is abstracted through metaphor.

In some cases, the difference that good metaphors can make in helping users understand a new product is enormous. When personal computers became popular in the late 1970s and early 1980s, access to computer functionality was mediated by an early operating system known as MS-DOS (Microsoft Disk Operating System). MS-DOS lacked a clear metaphor that might have helped people unfamiliar with the details of the operating system easily understand it. The problems that this fostered are reflected in the user interface and in the details of commands that, even today, can be challenging for all but the most knowledgeable users. Thus, the complexity of using these computers was a real barrier to public acceptance. It is no surprise then that early home computers were met with extreme skepticism. For example, in 1982 Pulitzer Prize-winning author William F. Buckley wrote:

> Some gadgets we know instinctively how to put to use: radios, say ... ... but a $1000 computer? The Pulitzer Prize belongs to the man who reveals what they're good for ... ... what they're good for that the average newspaper reader wants ... [1]

The semantic leap that brought mainstream use came when the metaphor of a desktop in an office was created on top of the "disk operating system." Users could interact with programs with clear visual metaphors in the form of icons that revealed their purpose. In turn they could "look at" those programs through windows, which afforded a view of their contents. The file system, previously a labyrinth of unintelligible names, became a set of folders that mapped to the filing cabinets already in offices. The operation of moving a file from one folder to another through a drag-and-drop gesture was easily understood by its analog in the physical world. There was even a trash can to throw things away. Together these metaphors helped make something that was previously scary and unknowable familiar and more friendly.

Digital Alienation

When users are faced with a new digital artifact, they have a problem. They need to make sense of the artifact from the perspective of their own semantic horizon; in Norman's terms, they need understanding.[2] Paul Ricoeur calls this task "appropriation," in the sense that it bridges historical and cultural distance, making one's own what was initially different (alien) and inaccessible.[3] This is a surprisingly common issue. For tasks as diverse as reading a manuscript from millennia ago and watching a video of a culture with which we have little contact, we must make "ours" what may seem to be alien. Ricoeur's notion of appropriation, which can be achieved through a process of interpretation, is the counterpart to the alienation that is intrinsic to such historical and cultural distance. Appropriation, in the hermeneutic sense, requires making sense of cultural and historical experiences distinct from our own. This is only possible by finding ways to relate such things to our personal knowledge and experience. Ricoeur summarizes this hermeneutic movement by saying that "interpretation brings together, equalizes, renders contemporary and similar. This goal is attained only insofar as interpretation actualizes the meaning of the text for the present reader."[4]

Digital technologies create a new form of alienation that we call technological alienation. A poem written in the sixteenth century exists in a different semantic horizon from our own, as does an Pacific island's myth of an autochthonous population. Similarly, the vast majority of people using mobile applications do not live in a world where the intricacies of software and hardware development are meaningful. The details of how data encryption works for wireless communication, or the minutiae of a smartphone's operating systems, form a horizon of meaning alien to most people, even those using mobile phones. This type of alienation may seem new because the technology is new, but it can be overcome by the same process we use to understand any part of our world that initially seems alien.

Thus, Ricoeur's proposed list of examples of historical and cultural "alienations" that require an interpretive struggle must be expanded. In today's world, it must also include technological alienation: the distance between the meaning of software algorithms, hardware components, and data structures, and the real-world experiences and knowledge of technology users. Of course, developers don't want users to see this process as a struggle, nor

Figure 3. The gulf of understanding. Communication between two people with vastly different horizons of meaning is virtually impossible. (Figure courtesy of Kira Chown)

do they want users to focus on what they do not understand; they want users to see things in terms that they *already* understand. To overcome the distance that might otherwise dominate users' perceptions when faced with new software, developers must create a metaphorical interpretation of their technology that feels familiar to users. This new reality should bring the developer's technical semantic horizon and the users' non-technical semantic horizon together, paving the way for users to smoothly overcome their technological alienation.

With technology there will always be a large gap between the expertise of developers and the knowledge of users. But how can one overcome this gap? How can one create a bridge between the complex universe of digital enablers and the world that users understand? This is the work of the first stage of a longer process, which we call the metaphorical cycle, that we continue to elaborate in the next chapter. Through the creation of an "impertinent predication" typical of metaphors, digital designers can describe

Figure 4. Metaphors are the normal way to bridge the gulf of understanding, allowing experts to communicate in ways that a lay person can understand. (Figure courtesy of Kira Chown)

enablers in a way that users can understand. This begins an often overlooked or minimized hermeneutic process that is crucial in the creation of a new technology. Descriptions of this creative process usually emphasizes the physical, mathematical, and logical achievements involved. Ironically, the public needs the metaphor, but celebrates the parts that they do not understand, probably precisely because they do not understand it—if it is hard to understand, then it must be hard to achieve. However, these enablers are literally meaningless to society until they can be brought across the divide of digital alienation so that the public can use them. This is inevitably done through digital metaphors.

Connecting Users to Technology

Not that long ago, most everyone lived in a predigital horizon of meaning. Our common cultural and traditional experiences meant that we had similar notions about what grocery stores, shopping carts, product aisles, sales assistants, cashiers, and receipts were. We inhabited, and still inhabit, these meanings. They are "semi-transparent" to us, easily understood and engaged. So too are the myriad associations that we have with these ideas. These are grist for comedians who do routines about the wheels on shopping carts that won't go straight or what it is like to stand in line. Such routines take advantage of our shared cultural experience in the same way that digital metaphors strive to do.

Digital designers, meanwhile, inhabit a parallel universe of meaning created through digital enablers. In this other universe, things like text boxes, buttons, scrollbars, sorting algorithms, network connections, programming languages, and web pages make just as much sense to the developers as a shopping cart does to us. Such enablers have meanings within that semantic space, and those developers have acquired those meanings through many hours of training and hard work. We could call this semantic horizon the digital one. Digital designers need to connect these nebulae of meanings, creating semantic bridges that link the predigital and the digital horizons of meaning. Further, they need to do so in a way that does not involve the same commitment of time and work from potential users as it did for them. We must be clear here, however, it is not the goal of developers to help users understand how text boxes, buttons, and the like work. The meanings that they want users to build are different from their own. Developers need to know how things like buttons work, but users simply need to know how to use them.

Thus, the first step in the digital metaphorical cycle happens when digital designers consider technological enablers and look for metaphorical ways to incorporate them into users' semantic universe. Here we return to the idea of metaphors as "seeing as." Designers can, for example, create semantic layers over the core of technological enablers. Such new layers might allow users to *see* what is really a machine learning recommendation algorithm *as* if it were an excellent customer service agent who points to other products in the retail store that the user is likely to want. Online store designers have created numerous metaphors to enable users to see icons, lists, buttons, recommendation algorithms, and communication protocols as if they were the shelves of the customers' favorite store. This familiar framework eases customers' move from the semantic universe of physical stores to that of digital purchases. Thus, a list of records in the computer's memory is displayed as images and text, seen by a customer as a shopping cart. Similar tricks allow entering credit card numbers into a set of text boxes and pressing a button to be seen as the familiar act of checking out. Thus, a common strategy for designers is to take what users are used to in the physical world and try to emulate it directly in software; they are composing metaphors to address new regions of meaning using digital ink.

Once digital metaphors are implemented in the form of hardware and software artifacts, they begin to interact and affect the universe of predigital meaning. Users need to figure out what they are seeing and how that is related to what they are familiar with, for example, putting an item in a shopping cart. They are invited (or guided) to see the pressing of a button as the physical pick-up of a grocery item and its physical displacement into the shopping cart. The impertinent predication "to press this button is to pick up the item and put it into a shopping cart" will require a new user to engage in a potentially difficult interpretative process to make sense of this new metaphor. One solution to this struggle pursued by some designers is called "skeuomorphism." The goal of skeuomorphic design is to make the digital metaphor as close to the physical experience as possible. For example, skeuomorphic book designs animate the act of turning a page right down to seeing the corner curl up. An early version of the calendar app for the iPad even appeared to have torn pages at the top, right under its fake leather, which, according to Apple engineers, was based on the leather in Steve Jobs's jet.[5] Even the icons of books in a digital library might have stitching designed to look like physical books, and those icons might be arranged on a fake bookshelf—some apps even simulate wood grain. Such designs are created to make the transition to the digital experience as easy as possible, but they have also been criticized for stifling innovation and for excessive visual clutter. In the early days of mobile technology, these discussions took place mainly on the internet; only recently have these issues begun to be adequately tested.[6] The best way to read a digital book, for example, may not be the same as the best way to read its physical counterpart. Digital books afford new possibilities for reading experiences that might be stifled by the very metaphors that are used to draw users in. We discuss this in more detail later in this chapter.

In *Transcoding the Digital*, Marianne van den Boomen investigates how metaphors are crucial in the way digital artifacts become used and understood.[7] For her, "the practice of digital code exchange can only be articulated, perceived, and conceived when it is translated into metaphors."[8] She uses a new media studies perspective to analyze the various levels at which metaphors mediate digital experiences, starting from the level of user interfaces and going all the way up to social discourse, with each level bringing important repercussions for culture and society. Van den Boomen's excellent description of how metaphors are essential parts of new digital media invites us to analyze the cognitive and semantic roots of this crucial aspect of digital technologies.

Digital technologies redescribe our experience in the world in order to be accepted and successfully integrated into the users' horizons of meaning. Digital technologies need to be redescribed in terms of everyday experiences in the world so that potential users can understand them. Similar to the way a Shakespearean metaphor invites us to see Juliet as the sun, a mobile application invites us to see a click on an icon as a token of friendship and appreciation. Thus, a mobile application can seek to redescribe trust and excellence as the number of stars users attribute to a ride service or a virtual store. Most digital artifacts accomplish this even while reflecting the tension between similarity and difference in metaphors. For instance, emojis are simple visual representations of emotions, but they are impoverished by the lack of immediate contact between the people communicating—a smiling face emoji does not convey as much information as an actual face. This metaphorical dimension of digital technologies requires an interpretative skill from developers and designers, a skill as important as any technical skill in the creation of reliable, efficient, and usable artifacts.

THE MATERIALITY OF DIGITAL METAPHORS

Metaphors are abstractions. As such it would seem that they do not exist in the world, and thus that they are not material. When someone is described as "climbing the ladder of success," there is no ladder and they are not literally climbing. Digital metaphors are different. They are implemented, and we interact with them directly. Further, they are "ostensive," meaning that they impose how they must be interacted with. These interactions, in turn, can be carefully engineered to guide us to understanding through their very use. This raises the question of whether digital metaphors are actually material.

Paul M. Leonardi explores this idea in "Digital Materiality? How Artifacts without Matter, Matter."[9] Most of us think of "material" as meaning that something is made up of matter. He suggests that this definition precludes us from calling digital artifacts material due to their lack of physical substance.[10] However, Leonardi's second definition of material, linked to what he calls "practical instantiation," is very close contextually to how digital metaphors work. For example, he notes that "justice" is an abstract or theoretical construct that gains materiality when put into laws (although without physical matter itself). The idea of value gains materiality through its practical instantiation in a monetary system as a particular kind of currency. In this sense, we can say that a similar abstract concept, "appreciation," can gain materiality through a digital metaphor. Many apps implement the metaphor as a heart icon that users can click when looking at pictures. Thus, the "I heart this" linguistic metaphor is instantiated in the form of this icon. The icon may be an artifact of software that is merely drawn on a piece of glass, but we see it as if it were real. Such practical instantiations are conspicuous in that they are repeatedly and explicitly shown every time an image is shared on a social network; we are constantly implicitly asked if we "heart" each photo. In addition, the metaphor's meaning is further reinforced by the consequences of clicking on a heart—invariably the app's recommendation engines will choose similar photos to show the user in the future. This is the model used by apps like TikTok, as well as the entire digital advertising industry. So, the interpretation of "I heart this" as "I want to see more of this" is ostensibly reinforced by the implementation, and thus the materiality, of the digital metaphor.

Literary metaphors, by contrast, exist only as words and are thus much more open to different interpretations than digital metaphors. This is because the pragmatism of a digital metaphor's implementation guides how it should be interpreted and provides feedback to the system. If the user does not interpret clicking the heart icon as liking the photo, then nothing in the context of the metaphor will make sense. But if they do interpret it correctly, they will engage with it. Whereas an app might react to such engagement, a poet is unlikely to have any idea how you have interpreted their work. In both cases, the digital or the poetic context guides the interpretation of the metaphor, but the context of a digital artifact includes its implementation. Further, the entire ecosystem of digital artifacts, the programs and devices that we have previous experience with, also guides and restricts the interpretation of the metaphor, helping make a "canonical" sense become quickly consolidated through the repetition that serves as the backbone of learning.

Bruno Latour discusses the implications of materiality using the metaphor "the speed bump is a sleeping policeman" as part of his investigation into how technologies affect action.[11] The metaphor provides a compelling example of why the relationship of materiality to the impact of metaphors is so critical. He states that "techniques have meaning, but they produce meaning via a special type of articulation that crosses the common-sense boundary between signs and things."[12] In other words, when we turn a linguistic "sign," such as a word or a phrase, into a material medium, like a physical speed bump or a heart icon, it affects the impact of a metaphor's agency in the world of its users. Driving over a speed bump brings the metaphor to life in a way that words may not. Thus, "arguments are wars" and "the speed bump is a policeman" take advantage of the same metaphorical mechanisms to produce meaning, but their ostensibility is different, which has important personal, interpersonal, and social consequences.

In turn, van den Boomen suggests that digital materiality raises the "operational reach" of digital metaphors.[13] The underlying metaphor of cell phone notifications describes blocks of text shown on the screen, and potentially the associated sounds and vibrations, as being reminders. The digital materiality of this metaphor in turn makes such reminders active and intrusive. The original source of the metaphor has mostly passive associations; previously one might have needed to look at the refrigerator door to remember that tomorrow there will be a sporting event for their child. This old-style reminder required us to actually look at the refrigerator or it was useless. On the other hand, the target of this metaphor, the smartphone notification, has a much more active set of associations, as digital reminders usually come with sounds, vibrations, and/or repetitions. Further, while we might only go near the refrigerator a few times a day, our phones are nearly always with us. Such digital reminders are designed specifically to remove the need for agency on the part of the user by transferring it directly to an app. With a reminder on a refrigerator, we had to hope that we looked at the refrigerator in the right way at the right time. With a digital reminder, the implementation can demand our attention, even overlaying and interrupting other things we might have been looking at on our phones. This is desirable, because we are less likely to miss appointments. Meanwhile, the repetition and intrusiveness also amplify the associative linkages in our minds. Thus, the metaphor's digital implementation makes its impact on users' agency especially strong, and its repetitions constantly reinforce that impact.

Van den Boomen and Hayles discuss the different ways in which Lakoff and Johnson's paradigm of conceptual metaphors needs to be expanded in light of such digital metaphors.[14] They propose that the expansion of the metaphorical effect should encompass not only the linguistic signs to which the metaphor refers, but also the materiality proper to digital objects and tools that embody such metaphors. So, not only are smartphone notifications metaphors to the linguistic sign "reminder," but this digital metaphor also embodies other nondiscursive attributes, such as the sounds, vibrations, and images in which the metaphor is expressed in its digital form. Van den Boomen[15] insightfully summarizes both the intrinsic metaphorical nature of digital artifacts and the transformation that digital materiality brings to the way metaphors impact experiences:

> Digital computer technology is extremely marked by metaphors. Here, metaphors nestle themselves not only in the representations of the technology and the discourse on its use and functions, but also in the technological objects themselves: the very thingness of digital objects consists of metaphors made material and operational. Such digital-material metaphors go beyond mere representation and language. They act as signs and metaphors, but also as things and procedures. The effects and implications of such sign-tool-object-metaphors are discursive and non-discursive, yet by all means material, embodied, and medium-specifically inscribed.[16]

Another example of the impact of digital materiality on metaphors is explored by Hayles in *Writing Machines*.[17] She investigates the implications of digital materiality for written texts. The digital medium brings a new form of materiality to texts which had previously been limited to paper. This new form has far-reaching consequences for the meanings expressed, actually or potentially, by the texts. On one hand, the new medium is mediated by metaphors. On the other hand, metaphors are shaped and constrained by their implementation in the digital medium. Digital readers, hypertext, and electronic-books (e-books) are examples of metaphors implemented as new forms of mediating written text. As such, they are also subject to the distortions of this new digital "incarnation." The materiality of digital artifacts creates new possibilities and constraints for the creation and expression of meanings, and it generates fundamental differences between literary and digital metaphors. As we have already seen, developers have choices about how much they push these new creations towards the purely digital, or how much they try to recreate their analog roots.

The e-book metaphor suggests that we see the images shown on the digital monitor as a paper book. In cognitive terms, the metaphor is intended to activate the same cognitive structures that we use when we read such books. The rectangular format, the organization of text in blocks that resemble pages, and sometimes even skeuomorphic effects with page-turning animations all create the hermeneutic and cognitive context for the metaphor to be understood and assimilated. All of these features work to prime and activate the familiar cognitive structure of reading that we have learned since we were young. Thus, a device that could be scary for some is made familiar and comfortable because it appears to be little different than an experience with which we are very familiar. However, the metaphor's digital materiality can grant new characteristics to the interaction with the text. For instance, the text can be easily "navigated" non-sequentially through hyperlinks. The digital materiality also allows the possibility that unknown terms can be immediately looked up in a digital dictionary, that segments of text highlighted by other users to be seen during reading, and a host of other things that are not possible with a paper book. There is a tension in cases like these between users who want to unlock the full potential of the technology and those who want to stick with what is comfortable. For those who use them, the new affordances of the digital medium may significantly impact the reading experience in relation to the metaphor's origin. Since our cognitive structures for reading books are active when reading e-books and the like, and since learning is always "on," reading such books automatically alters those structures. We will show in chapter six that our category for reading books is effectively split, with subcategories for digital and paper cases. For example, the category for digital books includes new features that were not part of our original category. The digital experience transforms the meanings linked with books as well as a whole constellation of concepts associated with the original category, such as what it means to buy books, to have a library, and to search for new titles that interest us. Once we get used to reading e-books, it is inevitable that, when we encounter an unknown word in a paper book, we can't help but bemoan the lack of a simple dictionary lookup. E-books may be a new category for us, but it shares a parent category with our category for paper books, since they are metaphorically linked. The e-book metaphor's digital materiality, as well as the designers' conscious choices to link it to "real" books, not only affects the meanings we give to such books, but impacts a vast network of cognitive associations, thus transforming a region of human experience.

It is important to highlight that the materiality of digital metaphors does not diminish in any way the fundamental impact that such metaphors have on the cognitive and semantic layers that we highlight throughout this book. On the contrary, the ostensibility of the digital medium amplifies the impact of such metaphors. It also accelerates the metaphorical life cycle, as we examine in the next chapter. The materiality of digital metaphors is a kind of accelerated metabolic process that ends up reducing the lifespan of metaphors, quickly incorporating them into the semantic horizon of their users and therefore causing them to vanish as productive impertinent predications. Returning to our digital notifications example, constant exposure and interaction to these unbidden messages popping up on smartphone screens quickly turns these digital metaphors into a new consolidated semantic unit. In point of fact, the predications of the metaphor's target (active, noisy) quickly override the typical predications of its source, paper reminders (passive, silent). As we will see in chapter six, learning is always on, building associative structures through contiguity and repetition. Thus, learning happens through the contiguity of the sounds, vibrations, and interruptions, timed to the appearance of the reminder. Meanwhile, due to the nature of the reminders, this is repeated time after time, day after day. As we build trust in these reminders, we use them more and more. Thus, between all of these aspects, such a system is perfectly constructed to maximize learning and to efficiently rewire our cognitive structure. This process is pushed even further when, as is the case with digital reminders, the new construct is apparently superior to what came before. We are then likely to increasingly rely on it.

METAPHORICAL STRUCTURES

Neither metaphors nor technology exist in isolation. The simplicity of a metaphorical statement hides the fact that metaphors rely on an accumulation of knowledge about the world. To understand that "time is money" requires knowing a lot about time and a lot about money. Similarly, apps in the digital world generally build and expand on what has come before. This is true in the technological sense, but also in the metaphorical sense. A new social media app can take advantage of both the existing digital infrastructure to ease its creation and the fact that users have probably assimilated a lot of metaphorical knowledge about social media, such as what it means to post something, to like something, etc.

The intricacy of the development of digital technologies and their associated metaphors becomes apparent as one recognizes their embeddedness in this multidimensional relationship of meanings and senses. So, a letter metaphor brings with it a set of associations (message, communication, post office, mail carriers, delivery time, privacy) that may enhance or complicate its understanding when applied to a technology.

As we shall see throughout this book, metaphors do more than establishing links between isolated words. At the language level, we will show how metaphors live in discourse and involve implicative complexes. Similarly, at the cognitive level, we will show how metaphors involve networks of associations activated by each part of the metaphor. This associative nature of the metaphorical process has significant repercussions for the creation of digital metaphors.

Just as the verses of the poem frame and guide the selection of associations evoked by a metaphor contained in the poem, the various components of a digital artifact (icons, sounds, colors, buttons, labels, gestures, screen transitions) need to be organized in a systematic way to guide its metaphor's resolution. For example, consider the metaphorical apparatus that has been created around sending digital messages. This apparatus includes open and closed letter icons, mailbox icons with a new message, the animation of a message icon moving quickly on the screen, the "send message" label, and the "mark as read" button, among many other things. In some digital mail systems, the button for composing a new message consists of a pencil and a piece of paper, a metaphor that is probably lost on many young people, but that is consistent with the overall bundle, a bundle which is trying to make a link in the user's mind between sending messages digitally and sending letters in the physical world. In fact, this linkage has been so successful that computer users retroactively renamed physical mail "snail mail" to emphasize its inferiority to the new digital incarnation.

The example further reinforces a central tenet of this book: that when we learn new conceptual entities based on metaphors, our learning is not isolated to that entity; because of its associative links, other concepts are affected as well. In many such cases, as with mail, the differences are used to diminish the original source of the metaphor. This may not seem consequential with regard to something like mail, but we will examine metaphors for larger concepts like conversations, relationships, and notions of identity. If we cannot help but compare email to physical mail, isn't it likely that we are also constantly comparing Facebook friendships to friendships in the nondigital world?

As in the example of mail, digital metaphors often operate in groups. Together they form semantic bundles that facilitate understanding complex functionalities and the expansion of new functionalities, e.g., the bundle surrounding messages includes the notion of a trash can for deleted messages, connecting it to the well-established trash can used by the "operating system is a desktop" metaphor. Similarly, functionalities related to digital security were organized to form an initial metaphorical bundle containing individual metaphors associated with comfortable concepts such as physical barriers and ways to invade such barriers: firewalls, passwords, keys, breaches, and attacks. Once users have apprehended such an initial cluster of metaphors, the job of introducing new functionalities into their semantic horizon is simplified. This can be done by using further metaphors associatively connected to the initial bundle, in this case keychains and Trojan horses. It is the guided construction of such a series of interrelated and mutually reinforcing metaphors that brings the predigital and digital semantic horizons together. Later, we detail some of the dangers of this approach.

THE LIFETIME OF A DIGITAL METAPHOR

In this chapter so far, for the sake of simplicity, we have mainly been considering a model in which metaphors connect a predigital semantic horizon to a digital one. This simplification has served us well, as it has helped us to explore some fundamental characteristics of digital metaphors. However, it obviously falls short as a description of what happens over time, since users incorporate multiple layers of digital metaphors through their everyday experiences with digital artifacts and thus build new layers on top of previous meanings in a hierarchy.

Thus, the "mouse" that is connected to our computers has become so common in everyday life that its metaphorical effects have effectively disappeared. The device's connection to the shape of the small rodent has become all but forgotten by the people using it. Thirty years ago, when people began using "mouse" as a metaphor for the digital positioning device, the associations of this linguistic sign were still strongly attached to the small rodent. Some users could (and this was the intention) even think: ahh, it looks like a "mouse." Nowadays, given the ubiquity of this device for everyday use and the fact that the only features shared by the device and the rodent are the relative size and shape, the metaphorical associations have largely vanished. Hardly anyone in an office would hesitate, unsure about what was meant, when a colleague asked to borrow their "mouse." The associative ties to the small rodent have been virtually extinguished over time thanks to the mechanics of the learning system that we discuss in chapter six. Meanwhile, new associations become connected to this metaphor now that it has been consolidated into its own category. Therefore, it was possible, and even easy, for tech companies to introduce the concept of a wireless mouse, even though the predication "wireless" is no longer related to the source of the original metaphor. Indeed, the original association of the device was based on the cord that connected it to the main computer, a cord that was reminiscent of a mouse's tail.

Ricoeur suggested that metaphors have a life cycle.[18] Metaphors are created and remain alive as long as their semantic novelty has not been fully consolidated into a new dictionary entry. While

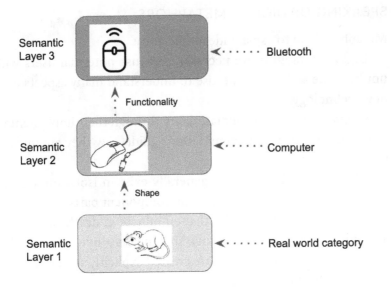

Figure 5. Growth of a semantic hierarchy over time. In Layer 1 we have an existing category. In Layer 2 with the introduction of a new technology the similarity in shape is used to create a new metaphor. In Layer 3 the addition of another technology is used to build on the metaphor even though the original metaphorical trigger is no longer in place.

alive, metaphors require effort to be understood; they remain open to new interpretations. They point to a semantic space that is not well defined. They are still a challenging impertinent predication. Once they are consolidated, metaphors lose their semantic flexibility and come to be understood like any other linguistic sign. The lexicon of a language is a graveyard full of metaphors.

While we will talk more about the peculiarities of this metaphorical process in the next chapter when comparing poetry and digital design, it is important at this point to highlight the idea that digital metaphors go through these life cycles and settle into new horizons of meaning within cultures and societies. Thus, the linear and simplified view of the simply binary transition from a horizon of predigital meanings to a digital one must, in fact, be understood as a more sophisticated and nuanced form of metaphorical spiral.

SPEAKING OF DIGITAL METAPHORS

Metaphors help to explain ideas efficiently.

Digital metaphors are necessary because potential users will not have the technical expertise to understand many aspects of a new technology.

Digital metaphors gain materiality through their implementations. In turn, these implementations guide how they are understood and used.

Digital metaphors do not generally exist in isolation, but in connected groups that reinforce the component parts.

Metaphors have life cycles that culminate in death when they are fully consolidated. For many, this means a new entry in the dictionary. Digital metaphors are created explicitly so that this life cycle is as short as possible.

CHAPTER FOUR

THE METAPHORICAL SPIRAL

In the previous chapter, we briefly introduced the notion of the metaphorical spiral and saw that digital metaphors have unique characteristics that separate them from normal metaphors. In this chapter, we discuss a particular variation on the metaphorical spiral as it applies to technology. Generally, the spiral is part of the broader ongoing cognitive-hermeneutic process of how we grapple with meaning and how we come at ideas at different times from different perspectives. In turn the notion of the spiral is supported by learning, meaning that with each encounter with an idea we are constantly updating our relevant cognitive structures.

Before moving on to digital metaphors, it might be helpful to step back for a moment and say a word about the spiral applied to nondigital metaphors. Think of the metaphorical spiral as a model that describes how metaphors change our horizons of meaning over time. Take, for example, when someone reads the metaphor "man is a wolf" for the first time. Before encountering this metaphor, they had a meaning for the word "man," which was part of their horizon of meaning—let's call this pre-metaphor moment their *prefigured horizon of meaning*. The exposure to the metaphor modifies this horizon; it alters the meaning of man by suggesting that we can see man as a wolf. The new sense proposed by the metaphor becomes part of their individual horizon of meaning and, if a community shares it, it becomes part of that group's horizon of meaning—let's call it their *refigured horizon of meaning*. Once integrated into a group's lexicon, they can suggest other metaphors. Someone sharing this new horizon of meaning might, for example, say that during the meeting, he showed his teeth, building on the common understanding that man is (or might be) a wolf.

It is only natural then, that digital metaphors are subject to these similar processes. There are, however, key differences that make a closer examination of the spiral in the context of digital metaphors a critical step in understanding how such metaphors are changing our world. One key difference is that when we normally engage with an artifact, such as a book or a film, we are in control. We can choose how and when we encounter it, and the artifact itself does not normally change, e.g., you will likely find watching the same film when you are thirty different than when you were a teenager, but the film itself is the same.[1] In turn, you might read articles or books about the film, or study the techniques that it uses, all of which will change what you think about the film, but the film still remains the same. All of this is part of the normal cognitive-hermeneutic cycle, reflecting how the meaning of things changes over time.

Things are different with digital technologies. Apps and devices are constantly changing, and app developers are thus active participants in the metaphorical spiral, which is a useful way of reflecting on the social implications of digital metaphors.[2] Mobile applications are efficient carriers and spreaders of digital metaphors due to the immense speed of their dissemination through app stores and over-the-air installation and update mechanisms. A metaphor such as "the number of stars reflects quality of service" as expressed in ride-sharing applications such as Uber and Lyft can quickly restructure the cognitive maps of users. Such users will then start to see these stars as a measurement of quality and might even apply the idea to other parts of their lives. The metaphorical spiral thus makes it evident how software metaphors have important social implications when redescribing horizons of meaning at the personal, interpersonal, and institutional levels.

Given the prominent role of digital technologies in modern life, it is essential to examine their social implications. Looking at the process of development through the lens of the cognitive-hermeneutic process will help us to do this without venturing into the intricacies and challenges of specific technology.

THE METAPHORICAL SPIRAL OF DIGITAL TECHNOLOGIES

Each cycle of this spiral involves the short life and eventual death of new digital metaphors as they are incorporated into users' semantic horizons. In turn, users provide feedback and new usage patterns that afford new directions for developers. The combination of the user's new horizon and their feedback can then be taken as starting points for the creation of new metaphors in other digital artifacts as the spiral continues. From a user's perspective, they begin with a prefigured semantic horizon. When they encounter a new technology or technological metaphor, that horizon is altered and reconfigured. In turn, developers respond to what users have been doing, and the cycle can continue.

The metaphorical spiral model can help us to analyze each level of this process. At any given time, users have a personal horizon of meaning which will necessarily include some metaphors that are already understood and incorporated into their understanding of the world. These prefigured metaphors can be both digital and nondigital. This is the users' world before encountering the new technology. In the second step of this cycle, a new digital metaphor is born as part of the development of a new artifact. This is the moment when digital designers attempt to bridge what users know and do not know with the new capabilities of the new prod-

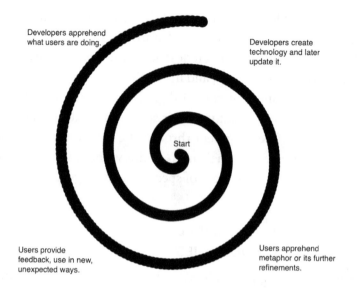

Developers apprehend
what users are doing.

Developers create
technology and later
update it.

Start

Users provide
feedback, use in new,
unexpected ways.

Users apprehend
metaphor or its further
refinements.

Figure 6. Metaphorical spiral of digital metaphors. The top half of the spiral corresponds to the developer's side, the bottom half to the user's side. This is a top view of a 3D process where looking into the image corresponds to looking backwards through time, hence the start point is the oldest in time. The outward expansion of the spiral signifies increased understanding and a deeper connection to meaning.

uct. The third moment of this metaphorical cycle occurs when these new digital metaphors connect with, and are incorporated into, the users' horizons of meaning. This is the moment of metaphorical refiguration, in which the user's interpretative process resolves the differences between their old model of the world and the new metaphorical view offered by the new product. Such a resolution generates a refiguration of meanings.

The metaphorical spiral model has the advantage of capturing how the interpretive processes of designers and users are intertwined in a sequence of moments. Moments in which the digital metaphors created by designers help users to incorporate new digital artifacts into their semantic horizons. When used, and therefore interpreted, by users, digital metaphors reshape the users' ways of seeing the world, their relationships, and themselves. The spiral also helps us understand that this is a repeating phenomenon with multiple subsequent cycles. A user's prefigured way of understanding and relating to the world is refigured by the new meanings proposed by the impertinent predications of the new digital metaphors created by designers and developers. Once appropriated by the user, the digital metaphor becomes part of how the user sees the world. It is integrated into their new prefiguration.

Developers

Developers encompass one half of the spiral. Given the predominantly market-driven ecosystem of digital technologies, developers' primary goal is typically to get as many users as possible to use their product. To do this, they must work in two directions. In the first, they work to make their product one that users will want to use. In the second, they work to ensure that users understand the product.

In terms of making products that users want to use, the digital world affords developers with unprecedented control. When users use digital products such as Facebook or TikTok their usage is fed back to developers in a torrent of data. Developers can see in real time how their product is being used and how changes in the product change user behavior. Thus, if a social network rolls out an interface change that is unpopular, they will know very quickly—and they can respond very quickly. Further, they can test out such changes by selectively rolling them out to some users and not others. Facebook even lets other entities use its A/B testing system.[3] Thus, users are a constant source of feedback and information that developers can incorporate into their own semantic horizons in order to keep those same users engaged.

As we have noted, none of this works if users do not understand the product. When developers put out a new product, or when they change an existing product, users need to know why they need the product or its changes and how to use them. Initially this is done by introducing digital metaphors, which are then modified over time as part of the cycle. Thus, Facebook's friendship metaphor changed when Facebook dropped its thumbs-up symbol in favor of a like button. A thumbs-up symbol will come with a somewhat different set of associations than a like symbol. In describing the change, Facebook noted that testing revealed that, on average, more people liked posts with the new like buttons.[4] In the same post, Facebook also noted how important the placement of the buttons was to driving engagement. Thus, even as Facebook chooses exactly how to present its metaphor, they too alter their users' cognitive structures.

The Facebook example provides a cautionary note. In this case, Facebook decided to change its product without feedback from users. It had the power to simply change its product. This is true for many digital technologies, even ones that are purchased by users. When you own a Windows PC, for example, Microsoft will update the operating system on a fairly regular basis. Many of these changes are to fix bugs, but sometimes these changes also change how the operating system functions, sometimes bringing new features and changing or removing old ones. Notably, some owners of these PCs may not want all of these changes, as they may remove or damage important functionality for those particular users. As Christopher Barnatt has noted, this is potentially worrisome as this behavior is normalized.[5] It is one thing for your PC to be updated in a way that you might not like, but it may be more dangerous with software like an autonomous car or a household robot. This example also shows how digital technology has altered the meaning of ownership. Once upon a time, ownership implied control. Owning things digitally often means that other entities can still exert a great deal of control over a thing that you "own."

Users

Nobel laureate Herb Simon noted as early as 1971 that attention is a precious commodity in an information-rich world.[6] In a world with so many choices, users need a good reason to devote their attention to a new app; this is often referred to as the "attention economy".[7] Thus, new apps must make sense from the get-go and build confidence as they are used, or they can be easily ignored or discarded. Users, therefore, want to answer questions such as "why do I want this" and "what can I do with it" very quickly. These questions are normally answered simply—"a cell phone allows you to have conversations with other people no matter where you are," or "you can easily stay connected with your friends, even at a distance." In turn, once the user engages with the technology, it begins to drive change in their understanding of the world.

Since apps compete with each other for attention, they have become exceptionally good at grabbing and holding our attention through a kind of evolutionary arms race. We will examine this in detail in chapter nine. The longer they hold our attention, the more they have the opportunity to work with our cognitive learning mechanisms. Thus, the more time we spend on a dating app, the more the connection between apps and dating is reinforced. The more time we spend having conversations by texting, the more the idea that texting is a form of conversation is reinforced, and so is the idea that being face-to-face with someone is not central to the idea of a conversation. In short, apps are very good at getting our attention, and once they have it, they automatically change our models of the world.

Users, however, do not need to be passive observers in this process. They are able to actively make choices. They can choose to use an app at all, or which one to use. And, notably, they can choose how to use those apps.

People are clever and creative, and they find ways to use digital technologies that may have never been anticipated by developers. One of the most significant early examples is the idea of smartphones as flashlights. Some astute developers realized that users had taken advantage of their screen's brightness to turn their smartphones into flashlights. One of the earliest hit apps in the iOS store, "Flashlight," literally just displayed a blank white screen. As a result, a technically elementary application that required no programming at all became extremely popular, making the best seller list at ninety-nine cents per download. Two other free flashlight apps were the two most downloaded utility apps of 2008. This unexpected use was so popular that today the most used operating systems, iOS and Android, already integrate the option of "flashlight" as one of its built-in digital metaphors, though they now use phones' built-in flash to do so. Scanning the technology news of the day will reveal numerous examples of users doing unanticipated things, from police playing Taylor Swift music while they are being filmed so the videos will be removed from YouTube for copyright violations, to children using soft drinks to get false positive COVID-19 tests, to interactive two-player games designed for AirPods where users each take one ear piece and then hear different stories.[8]

Edward Tenner and Margaret Morris have both explored the various ways in which the original configuration of digital metaphors can bring about completely unexpected consequences when they encounter users in the real world.[9] Morris, for example, comments on the case of mobile dating apps that initially offer metaphors for "flirting," "meeting," and "dating," but are seen by a subset of users as psychological mechanisms to cope with losses and promote emotional healing.

With the global distribution of mobile applications, these encounters with users' semantic horizons are even more diverse, given specific differences in each culture. For a programmer in Silicon Valley, it may be a simple matter to translate the language of an app for other countries, but it is a much more complex problem to account for such cultural differences.

In all of these ways, the meeting of digital metaphors and the potentially vast and potentially diverse pool of users will result in fast feedback and thus will begin the metaphorical spiral anew.

It is worth noting, however, that in many cases users may not feel like a part of this spiral and may feel individually like they do not have any agency. Markham, for example, in a study designed to prompt people to think of unusual potential futures for technologies, found that many people dismissed them out of hand. Markham distilled a typical comment as "It's just the way Facebook [fill in the blank platform here] works. Well, it's not like we can change Facebook, right?"[10] Such users see themselves not as active participants in a give-and-take with developers, but as passive observers forced to accept whatever the platform gives them.

ACCELERANTS

At its heart, this is a book about learning, meaning, and how those things interact with the digital world. One aspect of that world is an architecture, powered by the internet and enabled by ever more powerful devices, that is increasingly optimized to accelerate the cycles that we are discussing. Smartphones in particular have features that greatly leverage the impact of digital metaphors. First, due to the growth of wireless networks, such as cellular infrastructure and Wi-Fi connections, they can be used almost everywhere. Their ubiquity and mobility make them accessible at all times and places. According to Statista, in 2018 in the United States, users over 18 interacted with their devices an average of 215 minutes a day.[11] In addition, users accessed their smartphones an average of 52 times a day, according to Deloitte's 2018 Global Mobile Consumer Survey. Digital metaphors embedded into mobile applications are constantly reinforced by this intense use of smartphones. Adam Alter makes the case that smartphones become *irresistible* by leveraging cognitive addiction mechanisms.[12] This omnipresence and irresistibility have important repercussions for the life cycle of digital metaphors implemented in mobile applications. As we will see when we discuss learning in chapter six, the main drivers of learning are contiguity and repetition, often colloquially referred to as "fire together, wire together." The intensity of usage of smartphones and their apps ratchets up the learning process to unusually fast speeds. The new predications brought by digital metaphors are thus quickly incorporated into users' semantic horizons and lexicons.

Smartphones also have an incredible reach in contemporary society. According to Datareportal, there were more than five billion smartphone users in the world by 2021.[13] This ubiquity causes the metaphors created by mobile applications to quickly embed themselves in the way users communicate and refer to phenomena. Mainstream media helps spread new metaphors further because it reflects the impact they bring to society. Thus, a "tweet" becomes part of the daily news and expands the reach of metaphors beyond the direct interaction with the mobile applications that carry metaphors. Social media, which is only about two decades old, quickly introduced "like," "posts," and "timelines" as new ways of referring to the sharing of information between friends. Those terms have become a kind of way of measuring friendship.[14] As dana boyd suggests, the use of these sorts of things redefines the very concept of friendship.[15]

And finally, smartphones are not only digital artifacts. They are also platforms that enable us to deploy, locate, and use other digital artifacts, each of which carries its own new metaphors. So new mobile applications can be distributed extremely quickly. The distribution system of digital metaphors that affects thousands or millions of users overnight through mobile devices and app stores also dramatically accelerates the reinforcement of the new cognitive connections proposed by digital metaphors. This happens not only through the direct use of applications but also in discussions with other people who use such metaphors and in the media. Therefore, the metaphorical "story" of a social network application quickly entered into the common use of interpersonal communications and was transferred to digital and nondigital media, accelerating the metaphor's lexicalization cycle. To get a sense of the speed and scale of this phenomenon, consider TikTok. TikTok launched in the fall of 2016 and had 800 million active users three years later.[16] It had been downloaded 1 billion times by February of 2019, and 2 billion times a year later, and by the beginning of 2022 was on track to hit 1.5 billion users by mid-year.[17] By contrast, it is estimated that it took the telephone 50 years to reach 50 million users[18] and 75 years to reach 100 million users.[19] Meanwhile, TikTok users spend an average of 52 minutes a day on the app, which means that worldwide people are spending hundreds of millions of hours a day engaging with TikTok. TikTok can also evolve very quickly, averaging a new version more than once a week, with more than 60 new versions in 2019 alone. In an even more extreme version of this, according to Stripe CEO Patrick Collison, the company deployed 3,350 new versions of its API in 2020.[20] By contrast, car lines are typically redesigned on 4- to 6-year cycle, and tweaks to existing models only come once per year. TikTok can achieve these numbers because it exists on the back of mobile phones, the most successful consumer product in history, as well as the architecture of the information network. TikTok does not need to be manufactured, nor do users need to go to a store to get it. This speed with regard to updates also reflects a greatly sped up cognitive-hermeneutic cycle, as users are constantly adapting and responding to small changes, and developers are busily tracking how the changes are received.

DIGITAL BEGINNINGS

The hermeneutic spiral requires two things to begin: a product and a metaphor. As stated in the introduction to this book, our interests do not lie in creativity as it relates to new products, but the creation of digital metaphors is crucial to our story. Digital metaphors are able to take advantage of the same cognitive and semantic structures used for metaphors generally, but as we have seen already, there are crucial differences as well. Two of these differences are especially important at the time of the creation of a new digital metaphor. One goes to the imperative faced by developers that their products be easy to understand, and the other goes to the fact that developers are faced with explaining the unexplainable. So, developers must explain new, ever more complicated technologies in ways that the average person can easily grasp.

Digital Metaphors Require Short Lives

Ricoeur's conception of live metaphors can help us contrast the life cycle of poetic and technological metaphors.[21] A metaphor is live until it is transformed into just another dictionary entry. While it is still live, it is susceptible to new interpretations and, therefore, capable of gaining new meanings through changes in usage patterns. Metaphors, metaphorically, have a life cycle. They are created in an eruption of new meaning, born intriguing as impertinent predications. Over time, a metaphor's meaning may become fixed as a new entry for an existing word in the dictionary or as a completely new word. So, the computer "mouse" gave a new meaning to a word originally used mainly to describe small rodents. Mouse as metaphor died the day it became a new entry in our collective vocabulary (or Wikipedia, for that matter). Shakespeare's "time is a beggar," by contrast, remains defiant hundreds of years later, open to new interpretations. No single new word or polysemic entry for time or beggar in the vocabulary captures the meaning pointed to by this still-living metaphor.

While poetic metaphors can remain vital and open to different interpretations, technological metaphors are intended to be assimilated in the easiest and fastest way possible. The goal of a poet might be to use words to cause delight or to make the reader think about meaning. By contrast, the delight that innovators shoot for comes from a different source, the desire to use the new technology. In other words, in poetry the metaphor itself is a source of delight, whereas in technology the metaphor is merely a means to an end. The creators of technological metaphors want to speed up the metaphorical life cycle so that the possibilities of meaning converge to their intended functionality as fast as possible. Where a poet might take pleasure in having a reader wonder about a metaphor a week later, for an innovator this would be a failure. The impertinent predication must quickly shed its impertinence: "touching" shall become a synonym of "selecting," "texting" of conversation, "following" of "friendship," etc. As we have seen, the materiality of digital artifacts as metaphor and the rapid spread of software applications through app stores facilitate this cycle by pushing the use of these metaphors into large communities of users and communication media rapidly and extensively. In turn, frequent use of the technology provides the repetition that is the major driver of learning in all of those users. Every time they hit the "like" button on a post or have a conversation with a friend through text messages, users reinforce the new metaphor and speed its demise, reshaping what it means to have a conversation or to act like a friend.

Digital Metaphors Do Not Always Connect Two Well-Understood Concepts

There are other difficulties faced by digital innovators that are different from those facing poets. When Shakespeare metaphorically linked time and beggars, he built on two concepts that were already well established in the audience's minds. This, and the nature of his milieu, gave him the luxury to elegantly make his case through language and examples. Shakespeare was not suggesting that time and beggars were the same thing; instead, he was using aspects of beggars to make us look at time in a new way. In the digital realm, innovators often only have one side of the metaphor understood by the target audience, and they are essentially tasked with building the second from scratch by using the first. Whereas Shakespeare can take us through the features of beggars that highlight things about time, a digital retailer needs to rely on iconography to convince us that shopping online is the same as shopping in a store. A digital designer cannot fall back on showing us that thing X has commonalities with thing Y, so they have to convince us that thing X and thing Y are actually the same. This is made easier by knowing a user's intent, which in a retail context is to buy things. That intent can be used to create the metaphor and later to help the user to understand the experience. Digital designers are not trying to show us that their online store shares some features with "real" stores; they are working hard to convince us that online stores *are* real stores—only better, because they are more convenient.[22]

Such easy mappings are not always available. An example from the "experts'" horizon of meaning is the Unix command "grep," which stands for "global regular expression print." For Unix users, mastering the grep command is essential to productivity, but despite this, and despite the fact that Unix underlies mainstream operating systems like Linux and macOS, there is no version of grep in the drag-and-drop world of today's major operating systems. Once Windows, Apple, and Linux committed to the desktop metaphor, they also in many ways were bound to the limitations of the metaphor. Part of that commitment involves minimizing text and typing in favor of things that can be done with a mouse. Meanwhile, grep is a command that is completely about text: understanding how text is stored and processed on a computer. Power users can still use grep in today's operating systems thanks to specialized programs, but the vast majority of users have no idea that such things are even possible. Thus, the desktop metaphor was crucial in bringing computers to the general public, but at the same time it shields those users from the full power of the technology because those same users believe that the metaphor is the one true reality.

SPEAKING OF THE HERMENEUTIC CYCLE (OR METAPHORICAL SPIRAL)

Over time, each new encounter with an artifact deepens and enriches our understanding of the artifact. We refer to this process as the cognitive-hermeneutic cycle.

Digital metaphors add special elements to the normal cycle. In particular, digital artifacts can be ever changing because developers experience their own hermeneutic process with the technology.

Developers are constantly responding to feedback and making improvements to their technologies. Simultaneously, users are using the technologies in unexpected ways and providing feedback.

The feedback loop is accelerated by the infrastructure of digital technology, which makes every aspect of the cycle fast and easy.

PART II
〝〞〞〞〞〞〞〞

FOUNDATIONS

Now that we have laid out the major themes of the book, we turn our attention to our methods of analysis. The next section develops these methods by drawing on relevant work from philosophy and cognitive science. In turn, we describe a program for studying the impact of technology on meaning, starting with the creation of a metaphor to describe the technology to users and then continuing in an ongoing cycle where users provide a steady stream of feedback and developers a steady stream of updates to the technology. Together, these processes help to deepen the connection between the metaphorical use of the technology and how it connects to the ways people apprehend the world.

In the next chapter, we look at meaning through the lens of philosophy. We begin by exploring how the production of meaning is an essential feature of human experience. We argue that all our relations with the world, with others, and with ourselves are mediated by layers of meaning organized in cognitive structures that are in turn associated with linguistic symbols. These symbols take the forms of concepts and predications. We show how the interpretation of these meanings is a daily, inevitable, and continually changing activity—we are constantly re-engaging with meaning, updating and deepening our understanding. We then conclude our initial look at the meaning of meaning by thinking about the ways in which new meanings are created. In turn, this leads us to the phenomenon of semantic innovation—how new words and ideas are created through language—with the creation of new metaphors as its central driver.

In the following chapter, we examine many of the same issues through a completely different lens: the properties of human cognitive architecture. From this point of view, meaning comes to us in stages through learning. First, we must learn to build cognitive representations of the objects that comprise the world, as well as the features that comprise those objects. Next, we learn how these objects are connected and how they interact. The basis of all of these operations is association, which forms models of the world, called cognitive maps, through the strengthening and weakening of neural connections. This learning is always happening; every act of thinking is simultaneously an act of learning. The resulting structures are the analogues of the structures that we describe in the next chapter.

We conclude this second examination of meaning by reflecting on how these new connections are generated and the implications of this process. Importantly, this includes how new meanings must by nature come to us relatively slowly and conservatively under normal conditions. However, we show that language, especially through metaphor, is, in cognitive terms, an extremely efficient way of evoking and reusing existing cognitive structures to make sense of new concepts and ideas. This efficiency is why metaphor is so important to this book, as we argue that digital metaphors are orders of magnitude more efficient than most learning because they combine the inherent power of metaphor with the characteristics of the digital world. Together, they massively leverage key aspects of learning such as repetition.

For most of human history meaning has been primarily defined by a combination of our personal experience through learning and cultural transmission through schools, books, and the like. We argue that, even as digital technologies are changing our relationships to concepts like friendship and conversations, they too are an additional means of cultural transmission of ideas and are perhaps even supplanting some prior forms, just as digital technologies have disrupted so many industries. If, for example, your meaning of friendship has been largely shaped through social media, then it will be difficult for someone who does not digitally engage, or even someone who does but not on social media, to understand how you relate to people; notably, they may also simply not be capable of such relations as they require digital mediation.

CHAPTER FIVE

MEANING

THE MEANING OF MEANING

Our first order of business in this part is to discuss the concept that underpins everything in this book—meaning. A question so fundamental that it is recursive: what does it mean to mean? As Augustine said about time—which tends to be true about almost any complex concept analyzed from a philosophical lens—if no one asks us what meaning is, we know what it is; but "if I were desirous to explain it to one that should ask me, plainly I do not know."[1] We certainly have an operative notion of "meaning"; after all, we use the word every day. To appreciate how digital technologies change meanings, we make a brief detour through linguistic and philosophical paths in order to enable a deeper understanding of how digital technologies are relevant to how we attribute senses to our experiences in the world.

Educated, a memoir written by Tara Westover, was one of the most influential books of 2018 in the US.[2] It was prominent on lists of the best books of the year by the *Washington Post* and stayed on the bestseller list of the *New York Times* for more than two years. In the book, Westover recounts how she was raised in a Mormon survivalist home in rural Idaho. Her father believed that judgment day was coming soon and that the family should interact as little as possible with the health and education systems so that the government would not track them down. She and her brothers were raised in a closed community, which meant that their worldviews were shaped almost exclusively by their father. Almost everything a typical young American in an urban center would understand, such as the terms that shape most young people's daily experiences—school, teachers, grades, etc.—were unknown to Tara and her brothers. For them, schools and hospitals were places where there was a risk of being coopted or worse by the evil arms of public authorities. Then, when Tara turned seventeen and started interacting directly with the educational system for the first time, the world became totally different for her, upending many of her beliefs.

Westover's book is fascinating for several reasons, but we wanted to highlight a few aspects of her narrative that are directly related to our question on meaning. In her memoir, Westover reveals the process of transforming her worldview, the meanings she attributed to basic things in our everyday experiences such as learning and visiting doctors. The book also surprises us by showing how concepts like racism and the Holocaust may gain different meanings, ones outside social norms, thanks to her father's unique interpretation of biblical passages. Thus, even a simple glass of milk could become a forbidden and sinful thing based on perspectives that many readers would find foreign. Throughout the book, many readers might ask the question, "how is it possible for someone to see the world like that?" This question is more urgent and relevant than ever in a world of polarized societies and groups of people who simply cannot understand how it is possible that someone on the other side of the discussion or in the other political party can see the world in this or that way.

We believe that the answer to this question lies in the fascinating, almost mysterious, way in which we, human beings, relate to our daily experiences. We do not understand things in a direct and unmediated way as visual, auditory, and tactile stimuli reach our cognitive system. Our experiences, from the simplest to the most complicated, are always enveloped in layers of meanings that are deeply influenced by the cultural context we inhabit. What for one person is a place of learning and social coexistence, for another is a government trap designed to control our minds. A certain food can be perceived as a mere source of protein, even a symbol of wholesomeness, for one person. But for another person that same food can be a source of anxiety and repulsion due to the economic and industrial processes involved in its production and distribution. For yet another person the food could violate a religious precept and therefore be seen as a deplorable desecration of a transcendental relationship with the sacred.

From a purely logical-empiricist view that equates meaning to verifiable propositions, such discrepancies are anathema. However, if we were to ignore or downplay such discrepancies, we would lose sight of the fundamental fact that human beings guide their choices, actions, and values based on the meanings they attribute to their experiences and knowledge, and that such meanings are volatile, ever-changing, and shaped by cultural processes. The focus of our book is on one of these cultural processes, the distribution and use of digital technologies that, as we suggested in the introduction, are increasingly central to contemporary societies. If we want to pursue a deeper understanding of how digital technologies affect our lives, it is imperative to examine them as artifacts and processes that generate and transform meanings within a cultural framework.

Science fiction allows writers to explore ideas of how technological change can change meaning. For example, the popular *Back to the Future* movies, a tremendously successful trilogy that began in the 1980s, use time travel as a mechanism to show how technology impacts how characters see the world. In the first film, the hero, Marty McFly, goes back in time thirty years to 1955 and soon gets into trouble that ends with his archenemy, Biff Tannen. To avoid a confrontation with Biff, Marty runs through the streets of his home-town. He eventually bumps into two boys playing with a wooden box on wheels that is a sort of crude go-kart, a toy car for kids, but Marty sees something different. Where the boys see their toys as cars, Marty sees them as skateboards—a popular technology for kids in the '80s and a favorite of Marty's—and thus ideal to speed up his escape. The characters from the 1950s are amazed as he quickly transforms one of the boy's toys into a skateboard. The characters are in awe because that technology, as simple as it was, had not yet been incorporated into their universe of meaning. Another relevant example from the third film comes when the eccentric inventor of the time machine—Dr. Emmett Brown (or Doc)—travels back to the year 1885 and comments to the locals that people in the future will no longer use horses but instead something like "motorized carriages." The comment sends the Wild West lounge audience into loud laughter, and one of them asks if the people of the future will no longer have to walk. Doc replies that they will walk and run just for recreation or fun. One of the interlocutors replies, in complete disbelief, "Run for fun? What the hell type of fun is that?" prompting everyone to laugh even harder.

Doc and his friends see the world differently because technology has changed the meanings of fundamental things like walking, running, and getting around. "Motorized carriages" are not just things that we use and can fully understand once we have gained significant technical knowledge. Technologies are significant first and foremost because they change how we attribute meanings to our everyday experiences and relationships. But what are meanings? How are they created and incorporated into our worldviews? These are the basic questions that we discuss in this chapter.

We do not intend to propose a new semantic theory or even to make an extensive review of the many semantic theories that explore different aspects of meaning.[3] Instead, we adopt a working definition that we borrow from Paul Ricouer's *Interpretation Theory*:

By meaning or sense I here designate the propositional content, which I have just described, as the synthesis of two functions: the identification and the predication. It is not the event insofar as it is transient that we want to understand, but its meaning—the intertwining of noun and verb, to speak like Plato—insofar as it endures.[4]

There are multiple reasons to choose Ricoeur's definition as our starting point. First, he proposed a consistent and comprehensive theory of meaning that takes into account aspects of classical semantic theories such as those of Plato, Aristotle, and Frege, and placed them in dialogue with more contemporary theories like Austin, Searles, Saussure, Strawson, Russell, and Paul Grice. Ricoeur bridged the seemingly insurmountable gap between the so-called analytical and continental approaches to the philosophy of language. Second, Ricoeur looked at the ways meanings change over time through metaphors, just as we are doing in this book, and that work was one of our sources of inspiration. Third, Ricoeur's theory places us in the context of a hermeneutical approach to digital technologies that will serve as a guide for our analysis of various technologies.

Since this book is not aimed primarily at philosophers or linguists, we need to unpack Ricoeur's statement a little bit to make it more understandable and to connect it to our hypothesis and goal. Ricoeur begins with the idea that the propositional content in a sentence can be broken down into two important pieces: 1) singular identification and 2) universal predication.[5] These pieces translate roughly to the subject and predicates children used to learn in school. When we talk about the meaning of something, it can always be expressed as a combination of the identification of something specific that is the subject of the sentence, and the predications that are the attributes assigned to that subject. These predications can be a class of things, a quality, a relationship, or a type of action. The sentence "Rio de Janeiro is the most beautiful city in Brazil" identifies a specific place and predicates it with a set of relevant attributes, in this case the country in which it is located and its aesthetically pleasing geography. Thus, finding the meaning of a sentence involves both identifying the subject and assigning predications to it.

While this first analysis of meaning is useful in that it provides us with a clear way of thinking about the non-contextual aspects of the-meaning-of-meaning, it runs the risk of reducing our analysis to being nothing more than grammar, an examination of linguistic structures devoid of any connection to the social reality in which the sentences were created. A crucial complementary way of thinking about meaning is to approach it from the perspective of someone who is not sure about a specific meaning. When someone asks you about the meaning of something, your answer will usually involve unpacking the various predications you intended for that particular statement. For example, when asked, "what does Rio mean to you?", one could reply that it is "the city of contrasts." Such an explanation could go further, exploring other possible predications of Rio: it is uniquely located between the warm South Atlantic and gorgeous ranges of mountains; it is the former capital of Brazil; one-third of its population lives in slums, etc. Notice that sometimes possible predications will be understood differently for the speaker and the listener, e.g., not everyone knows that Rio was once the capital of Brazil, thus complicating the issue of whether a sentence can have the same meaning for two different people. Placing the question of meaning in the context of the use of meaningful sentences increases its scope, connecting it with the temporal fluidity of history, and adds layers of complexity not possible with more static approaches to analyzing predications. It invites us to consider extra-linguistic implications of meaning and the relevance of such implications in our daily lives.

INHABITING MEANINGFUL WORLDS

Meanings Matter

As discussed in the previous section, meanings are much more than mere linguistic phenomena. They alter how we understand and evaluate the world and ourselves. They impact our actions and our life plans. If we predicated "Rio de Janeiro" as "the most naturally beautiful city in South America," we might feel tempted to prioritize our savings in order to buy a travel package there. On the other hand, if we look at Rio as "a city of contrasts" or "a city with a high crime rate," we might redirect our savings towards an extended visit to Tokyo or Sydney. The question of where to take our vacation is certainly made more complex because of the set of predications that work together to collectively build up the meaning of something. Further, these predications change over time, depending on shifting social, cultural, and technological contexts. The bottom line is that we understand, evaluate, and relate to things, other people, and ourselves through the lens of meanings, and those meanings rarely stay fixed.

Filters through which We Experience the World

In his book *An Essay on Man*, Ernst Cassirer, an influential twentieth-century philosopher, wrote that meanings (particularly those he calls symbols) operate as filters between humans' receiving and operating systems, and that this mechanism for filtering our sensory input is a key distinguishing feature between humans and other species.[6] Thus, a hug is never perceived just in terms of the excitement of different nerve endings in the peripheral nervous system. There is always a symbolic dimension that makes such tactile excitement relevant and meaningful. The symbolic system adds predications to physical sensations, making the hug "a manifestation of affection," or "a gesture of reconciliation," and so on.

For Cassirer, these symbolic filters are so relevant that they justify calling men "symbolic animals."[7] For him, being human is about more than just being rational. Many other species are able to use what we could call reason: creating models from the stimuli they receive from the environment and adapting their behaviors based on these models. Cassirer claims that the filters of meanings that we place between what we receive through our sensory organs and our reactions is what distinguishes the way of being proper to humans.

Given the importance and thus the ubiquity of this phenomenon, we could spend the rest of the book exploring examples of these symbolic filters and how they work. But since we still want to keep with the same book title and reserve the bulk of the text for digital metaphors, we choose four especially salient areas of our basic bodily and social experiences that reflect different levels at which these symbols occur: to eat, to mate, to die, and to communicate. These are also areas that have been dramatically impacted by digital technologies and the transformation of meaning such technologies foster.

"To Eat"

Our first example involves a aspect of our bodily subsistence. When we see an object, a predication related to the edibility of that object immediately presents itself: *"That is not only a vaguely spherical red object that I know to be called an 'apple,' but it is also something I can eat."* Such predications are dependent on the cultural and historical environments that frame our experience with objects and animals. If you have traveled to countries with different cultures from your own, you have probably come across things to eat, typically vegetables or animals, that you may find personally repellent. For many people, reflecting many cultures, eating things like chicken feet, sheep's heart, haggis (minced liver and lungs mixed with onions, oatmeal, and suet and seasoned with salt and spices, cooked inside the animal's stomach), or some insects like grasshoppers and scorpions would be unthinkable. For other members of the same human species, these items are not only edible but are considered delicacies. The critical aspect underlying such a vast range of behaviors towards the same object is that humans do not experience the external stimuli immediately—symbolic filters learned over time, through cultural interactions, always mediate them.

But the symbolic layer of meanings on top of our eating habits goes far beyond the attribution of edibility. Consider, for example, that some types of food have become attached to specific festivities, or that some foods have dimensions of sacredness in some cultures but not others, and how meals organize the daily rhythm of life in several contemporary societies.[8]

As we will see in each of these four core areas of our daily lives, technologies profoundly impact the meanings we associate with food, from the mechanisms used to generate fire, to the utensils used to eat, to the furniture designed for food,[9] to the advanced appliances found in some current societies' kitchens. Digital technologies go a step further, fundamentally changing how we choose what to eat, how we buy food through shopping apps,[10] how we prepare food through recipe apps, and how we share dining experiences with friends and family. It was a widespread practice during the covid pandemic, for example, for families to meet through video-conference applications, such as Zoom, and share celebratory meals on special occasions remotely. Many other people simply post pictures of special meals on social media.

Long before the COVID-19 pandemic triggered a boom in the use of online delivery apps and grocery shopping, a 2013 article emphatically suggested that "touch screens are becoming as integral to the restaurant experience as knives and forks."[11] What we eat was mediated by the experience we had through websites and mobile applications. An enticing screen layout, an easier table reservation functionality, and ominous rating mechanisms reshape how we select our meals and envelop our biological need for nutrition with yet another transformative layer of symbolic mediations.

"To Mate"

Our second example exploring symbolic mediations involves finding our mates. At the purely sensory level, the pairing process involves physical stimuli and aims for the highest rate of successful reproduction possible to guarantee the maintenance and eventual increase of the species.[12] However, a description at this level is not only unusual; it may sound either rude or profane to many of us when it comes to discussing our own species. A significant part of our societies' symbolic layer concerns how we structure our sexual relations—what is allowed and what is not in terms of gender pairing, how gender is defined, how individuals maintain their sexual partnerships, and the consequences of eventual voluntary or involuntary disruptions of such pairings. Consider, for example, the meanings surrounding the pairing of individuals. In some societies, marriage represents the culmination of these meanings. This idea brings with it an immense apparatus of associated meanings: the initial meeting of a potentially romantic couple, the experience of dating, introductions to friends and family, planning a wedding ceremony, the ceremony itself, monogamous or polygamous family structures, laws surrounding the partition of goods and breaking of nuptial contracts, and more. It is easy to imagine or remember narratives, movies, books, etc., in which any or all parts of this process are the central focus.

If we pause for a moment to think about the layers of social meanings connected to "to mate," it becomes clear how ubiquitous they are. Precisely for that reason, they have become transparent in our daily experiences. Cassirer tells us that we live immersed in this symbolic universe and we cannot escape it.[13] Thus, it is no longer possible for us to think of a hug as simply a set of tactile perceptions in the same way that it is no longer possible to think of a large gray four-legged animal with big ears and a trunk as anything other than an elephant.

In recent years, the world of romance has been increasingly impacted by digital technologies. For instance, encounters involving potential romantic partners are increasingly mediated by dating apps. There were an estimated 44.2 million smartphone dating app users in the United States in 2020.[14] This means that the series of social conventions that have long mediated the selection of new partners is being profoundly reframed. Meeting someone once meant being at the same place at the same time, thus generally limiting, for example, potential partners to relatively small communities. This is no longer the case. Relationships can be established virtually and can stay that way for long periods of time.[15] Digitally mediated flirting also means that there are typically fewer emotional risks and less exposure involved in the decision to propose a possible romantic partnership.[16] "Courting" is moving online just like everything else. Digital technologies can also be used to facilitate fleeting encounters in physical environments, such as bars and clubs, and as a way to extend or deepen such relationships, creating new combinations of physical and digital interactions between partners.[17]

Some groups, such as seniors, have begun to recognize the possibilities of completely reframing their lives based on their ability to find partners through digital platforms. For them, the nourishment and maintenance of emotional relationships can also become mediated by new symbols and meanings brought by digital technologies. For seniors who might otherwise have been consigned to long stretches alone, this can be a godsend. In particular, messaging applications have become the primary mechanism for maintaining links between partners. This has brought with it another new set of mediations. Thus, the time to reply to a message might be associated with a potential partner's interest in the relationship. For example, a delayed response may signal that there is a problem in the relationship, triggering anxiety and worry. Texting conventions such as emojis and abbreviations may also add new layers of interpretation. Consider the simple example of responding "okay" to a text. In the world of texting, there are four distinct responses: "okay," "ok," "k," and "kk," which all mean different things. For example, "k" is considered to be rude by young people, whereas "kk" is very agreeable.

"To Die"

The end of individual physical existence is often recognized as one of the few constants in life (the other being taxes) and, as far as we know, is a common trait among all species. Most other species do not seem to have particular behaviors associated with the death of an individual, although some have their behavior mediated by chemical reflexes evolved to preserve the group's well-being. However, several species do mourn the death of their individuals.[18] This phenomenon is well documented in elephants and chimpanzees, for instance, and extends to several other species. Nevertheless, the human symbolic apparatus goes far beyond what we know about the reactions, even sophisticated ones, to an individual's death in other species.

Almost every religion has a series of symbols and rituals associated with death. One common mediation around death, the concept of an immortal soul, is paradigmatic for grasping the potential power and influence of the symbolic universe. This concept can, and has, effectively created new meanings around existence for billions of individuals. It operates fundamentally in the symbolic layer; it is not something that we are capable of directly perceiving. Our observation of this phenomenon does not imply any ontological assumptions or consequences. We are not making any assertions about the reality of a soul; instead we are highlighting an experience that is central to billions of individuals, people living within our symbolic systems. It is worth noting that such mediations are not limited to reframing the end of physical existence. Consider, for example, the metaphor that "death is a passage." Symbolic mediations such as this affect how people live, make choices, and set priorities in their lives. To give just a few examples, Buddhist monks retire in monasteries for meditation practices, extremists of various religions commit suicide to perform deeds in order to pave the way to a good life after death, and ancient Egyptians embalmed bodies and secured physical supplies for the afterlife journey. In these and many other ways, the symbolic layer around "to die" has far-reaching and decisive consequences for what it means "to live."

Recent advances in technology affect the meanings associated with death in ways that are sometimes overlooked. On one hand, recent advances in therapeutic techniques, more sophisticated forms of treatments, and vaccines have decreased the number of early deaths, particularly infant mortality. In total, these advances are perceived almost without exception as positive. On the other hand, even as early death has become rarer, and as the transcendental meanings attached to it are being lost as religion has declined as a force in some societies, "to die" has come to have an even more threatening set of meanings. Since early death happens less often, it becomes more important when it does. Thus, many contemporary societies tend to minimize any mention of death; in some societies death is perceived as something grotesque and even offensive. Advances in genetic engineering have also brought new perspectives to the extension of human life, and studies on the genetic mechanisms of aging have opened doors to new frontiers of longevity. These and other changes brought about by new technologies impact, challenge, and restructure various meanings around "to die."

Digital technologies have added yet more layers of meaning to death. First, such technologies dramatically changed how death is communicated. When a loved one dies, it is easy to inform a global network of people about it, and they can respond in real time. Moreover, the concept of digital immortality, something that until a few years ago was a work of science fiction, is being explored by some digital artifacts. These artifacts collect digital traces from individuals, such as emails, posts, videos, and messages, in order to find communication patterns. In turn these machine-learned patterns allow these artifacts to simulate communication with the deceased individual, even after their death. Several dead music artists have given concerts or even gone on tour.[19] In January 2021, Microsoft received a patent for software that could represent people (including dead ones) as chatbots.[20]

"To Communicate"

Our brief journey through some of the more important corners of the human symbolic universe concludes with the idea of "to communicate." While several species have developed sophisticated and efficient communication mechanisms through signals encoded in sounds, movements, and odors, human beings are unparalleled in our ability to create complex linguistic systems. These allow us to tell stories, develop arguments, and even create new concepts. Based on incredibly malleable and adaptable linguistic systems, human beings have been able to build and expand our symbolic universe to the point that this mediation layer is ubiquitous, helping define our perceptions and actions in the world.

At this point, it should come as no surprise that technologies have impacted the meanings of "to communicate," beginning with the invention of writing. Through the written word, it became possible to communicate not only with contemporaries, but also with individuals who did not yet exist. Writing thus changed the temporal aspect of "to communicate" so that it was no longer limited to a singular historical moment. With writing, one could meaningfully say "this message is for the generations that follow (and it will be accessed with no need of intermediary messengers)."

Writing, coupled with advances in transportation, transformed "to communicate" to also mean communicating with people in other locations, including different countries and continents. An immigrant in Argentina could finally communicate with their family in Italy. Such a thing simply would not make sense, for example, to a European in the Middle Ages for whom the meaning of communication was intrinsically linked to the local community, their village, or fief.

If the combination of writing and advances in transportation have changed the meaning of "to communicate" to include communication over vast distances, communication to a broad audience, and communication aimed at future generations, then digital technologies have further made such communication instantaneous, regardless of distance. And with that, technology has altered almost the entire semantic context of "to communicate." It is now possible to communicate with someone in the Netherlands, Florida, and Japan simultaneously with just one video call. A friend, wherever they might be, is always a couple of touches on your smartphone away. Such changes dramatically affect our symbolic universe and, therefore, how we live and experience the world. We revisit this mediation in more depth in later chapters.

Homo Interpretans

The immense richness of meanings of the symbolic universe demands interpretation, and the enormous wealth of predications intrinsic to such mediations creates infinite possibilities for meaning. But the malleability of the symbolic universe comes with a price—ambiguity. Such richness in meanings requires a commensurate and constant effort to understand which predications are currently appropriate for each symbol we interact with. We are constantly exposed to the arduous task of determining the most appropriate meaning for a given linguistic or nonlinguistic sign. A handshake can mean a farewell or a reconciliation. A smile can mean contentment or subtle and polite disapproval. And the cost of misinterpreting such gestures can be severe. When we ask our partner, "Did you sleep well?" they may interpret this set or linguistics signs (words) as "He's concerned for my well-being," or, conversely, "I must look terrible this morning for him to ask this." As you might well know if you have been in this situation, the particular interpretation chosen by your partner may have critical implications for you during that day (if not longer).

This is even more evident in the linguistic spectrum. The same word or expression can have different meanings depending on the context and intentionality in which they are used. Autoantonyms are not only an example of a linguistic sign with two different meanings, but they represent the extreme case in which the same sign (word) has contrary meanings. For example, in English "to clip" can mean "to attach" or "to cut off"; and oversight can mean "accidental omission or error," or "close scrutiny and control." In the next chapter we'll see that even identically worded sentences can have two different meanings depending on contextual clues.

Because we are always interacting with the world, and with others, through symbolic mediations and because these mediations are often polysemic—they can have different meanings—we constantly need to interpret the meaning of what we see, hear, and read. In other words, the consequence of living in a world mediated by symbols with different, ever shifting, and possibly conflicting meanings is that the act of finding the appropriate meanings for a given sign is a core element of being human. Johann Michel, in his recent book of the same name, calls human beings *Homo interpretans*, which provides an interesting complement to Cassirer's suggestion, *Homo symbolicum*. If we create a symbolic world with countless explicit or latent meanings in each action, gesture, or speech, we also must constantly interpret these meanings.

Modes of Interpretation and Evolving Meanings

As Michel points out, the levels at which these interpretations occur vary in intensity and depth. Some interpretations of meanings are built in to our genetic codes, so when we see something that could mean a risk to our lives, we run away immediately. This is what he calls proto-interpretation.[21] In the next chapter we discuss proto-interpretation in the context of how attention works. Then there are daily interpretations, like a morning greeting or a news story we read on the internet. Finally, we have a level at which we think about the fact that we are interpretive beings, what interpretations are, and how they affect us; Michel calls this level meta-interpretive (in fact, it is exactly at this level that we operate throughout this chapter).

Up to this point, we have significantly simplified our discussion and only marginally considered how meanings change over time. Something that in Michel's model could initially be classified as "interpretation" becomes "proto-interpretation" with use and habituation. Traffic signs are good examples of how the need to interpret some symbols consciously passes over time in a subliminal interpretive process.

Therefore, we have to expand our analysis and complement it with a dynamic perspective that integrates the process of how meanings are linked to our world experiences. Imagine, for example, that you want to understand a new sport. This was recently the case for one of the authors of this book who, being Brazilian, had very little contact with American football during his youth. The very fact that a sport that primarily uses the hands is called football provided an immediate interpretive challenge. Thus, in the first contact with American football, there is a set of meanings and symbols that must be deciphered. The referee's gestures, the movement of the players, the strategies of a play, and the reactions of the public are all full of meanings that require a conscious and constant act of interpretation on the part of any new fan. However, after a few games and then seasons, these meanings are increasingly incorporated into the new fan's meaning horizon, and the interpretive effort becomes different. At this point many predications have been consolidated and can be interpreted with a greater degree of naturalness—almost automatically and effortlessly.

Without the new fan even noticing, terms like touchdown, interception, and tackle transition from confusing to familiar to old hat. More than that, these new meanings become part and parcel of how the fan thinks about the world. Suddenly, when one of his children achieves something important, he might think "touchdown." He has not only learned some typical predications of the concept of touchdown, such as "achieving a goal by overcoming barriers," but he actively applies this meaning in other contexts. This process of actively interpreting new meanings, and the corresponding gradual reinforcement and internalization of these meanings as they become familiar, will be central to our discussion in the next section.

EXPLORING NEW WORLDS: SEMANTIC INNOVATION

The meanings and concepts that shape our goals and priorities and mediate how we see reality are organized in a network of semantic predications (X is Y), as we discuss in Appendix A. These networks are created by interconnecting individual predications of things, experiences, people, and actions. As such, our internal semantic networks integrate various concepts while building a complex and multifaceted structure that constitutes how we understand our experience in the world. In this section we refer to this structure as our "life-world."[22] For example, in one person's life-world some predications of Rio will be connected to concepts centered on nature, reflecting Rio's natural beauty, while others will be linked to the concept of a big city. In turn, "city" and "nature" have their own network of predications that may be evoked while trying to make sense of Rio.

To complicate things further, meanings are not static, neither at a personal nor at a social level. Concepts and predications in our life-worlds are always being created, always changing, and often used in communication. This is happening constantly through learning, social interactions, scientific discoveries, and changes in the world. For instance, Rio de Janeiro hosted the summer Olympic Games in 2016, creating a new predication for Rio in the life-worlds of people across the globe. In turn, people watched, read, and talked about the games, building their networks even more.

A sports fan who ventured to Rio for a visit during the games would have been bombarded by new predications of the city as they were exposed to new direct or indirect experiences of Rio. From the moment they started reading more about Rio as they prepared for the trip, or engaged in conversations with friends who visited Brazil before about what to expect from the city, their predications were being expanded and altered. And when the actual trip happened, it would have completely reshaped their previous predications about Rio. Each sensory experience during their stay in the city would serve to expand or alter their predications of Rio de Janeiro, changing its meaning radically.

Our life-world is constantly changing, either because things we are familiar with are actually changing or because we are experiencing new things or different aspects of known things that did not change substantially. One does not need to travel to another continent to change one's life-world. When one meets a new person, the predications that are suitable for that person are still unknown. Getting to know someone or something is a process of choosing which predications we judge appropriate for that person or thing. By making such choices, we gradually define what that person or thing means to us. "She is clever," "he is tall," or "they are just."

If there is one constant in our day-to-day experiences, it is that we are always building and expanding meanings. When we learn something new, we build new meaning by using a series of predications to make sense of that experience. Anyone who has spent time with small children has seen this in action as a child's repertoire of meanings grows exponentially to account for the immensity of new things, people, actions, and situations that need to be integrated into their life-world. However, this constant attribution of meanings does not stop in childhood; it extends throughout our lives. This is true for subjective predications (we perceive new things about our experiences in the world) and for objective aspects (the world around us changes and we need to adapt the predications that we attribute to it). But how do we create and update meanings to help make sense of our ever-changing worldly experiences? Answering this crucial question is the main objective of this section.

From Action to Language and Back

In the previous section, we navigated from language to action in the world. We saw how the meanings we find at the linguistic level are mediators of our choices and actions, shaping our life-world. In this section we return to language to investigate an important way of creating new meanings—metaphors. Towards the end of the section, the exploration of the linguistic model will once again move us from language back towards our real-world experiences. We examine how metaphors are a privileged form for creating new meanings—semantic innovation. They aid us in exploring new meanings by creating new concepts that challenge and expand our current life-world. We argue that metaphors are not merely "privileged" ways to expand our meanings, but in many ways they are the predominant way. This is due to two of their core features: 1) they speed up meaning creation, and 2) they have the unique capability of providing meaning to some experiences and phenomena that have not already been captured by language. Let's look at each of these characteristics in turn and unpack them with examples that will guide us to the ways in which metaphors, as the cornerstone of semantic innovation, are critical to digital technologies.

Do You Know John Doe?

Consider the usual situation of meeting someone, John Doe, for the very first time having no previous knowledge about them. To "make sense" of a new acquaintance, we have suggested that one learns predications that are either gained through first-hand experiences or by being told about the person by others: "He is hardworking." "He is gentle." "He works at a research institution." . . . In either case the new predications come fairly slowly, one at a time. We discuss the mechanics of this process and why it is true in the next chapter. However, language offers another way to create new meanings, one that is more efficient—by using creative comparisons. Such comparisons can speed up the process of making sense of something by transferring well-established predications from one thing to another. For example, if a friend says, "John Doe is a bee'," they are suggesting that we see John Doe as a bee, and therefore that we can leverage the usual predications associated with bees to begin to understand John Doe.

Certainly, metaphors are not the only mechanisms for assigning meaning through new predications, but they are extremely efficient because they allow the concise association of a large set of assignments from one term to another within the metaphorical statement. In saying that John Doe is a bee, we transfer a group of bee predications to John Doe, such as "John Doe is productive, just like bees'," "John Doe is reliable, just like bees," "John Doe is hardworking, just like bees," "John Doe is collaborative, just like bees." We note here, as everywhere, that such predications are steeped in local culture and experience and thus may vary. Fans of Mohammad Ali might wonder, for example, if John Doe "stings like a bee."

Expansion of Meaning

By bringing two concepts together, metaphors facilitate and accelerate the process of interpreting complex phenomena. Take a visit to a doctor (assuming you are not one yourself). Doctors often need to communicate the meaning of complex biological or biochemical concepts in a way that is accessible to lay people. So, they might offer the metaphor that our immune system is an army tasked with fighting attackers that are hostile to our body. For your part, you might start to see your immune system that way, and that new point of view will change the meaning you give to a fundamental thing in our human experience—disease. A mother might tell her son after a visit to our metaphorical doctor "you need to eat nutritious food so that the soldiers in your immune system are strong and ready for battle."

This ability of metaphors to create meaning for new or complex experiences is particularly important in making scientific and technical advances accessible to consumers. Let's look at our example from *Back to the Future* again. In the third film, Dr. Emmett Brown (Doc) is back in 1885 and comments to the locals that people in the future will no longer ride horses and instead will use something like "motorized carriages." In other words, to explain the meanings of the new technology that people have no concept of, he uses a metaphor to put his explanation in terms that they already understand. He used the known concepts of motors and carriages to explain something new, something that still did not exist.

But metaphors are more than a way of transporting predications from one concept to another. They can also expand and redefine meaning in innovative ways, affording the ability to create completely original concepts. These new concepts can then be used to make sense of a particular phenomenon. Max Black, analyzing the components of scientific knowledge, said that "perhaps every science must start from metaphor to end with algebra; and perhaps without metaphor there would never have been any algebra."[23] He insisted the metaphors and imagination were critical components of science and humanities.[24] Wyatt has also commented on how metaphors explore the future of the internet and other digital technologies through imaginaries, not only for technologists and engineers, but also by policymakers, journalists, academics, and industry spokespeople.[25]

In the introduction, we discussed the "paintbrush as a pump" metaphors proposed by Schön. He called this a generative metaphor, since it provided a new source of ideas for a research team that was struggling to optimize the performance of the paintbrush. Schön realized that "paintbrush-as-pump was a generative metaphor for the researchers in the sense that it generated new perceptions, explanations, and inventions."[26] Generative metaphors are examples of metaphors that not only communicate concepts and ideas already organized and established through other logical and formal procedures, but that also may provide new ways of thinking about those concepts. Metaphors are also seminal, creating new perspectives to explore phenomena's aspects and characteristics that are not accessible with the concepts already established in the language.

Meanings Grow Metaphorically

Imagine that you are back from your exciting first visit to Rio de Janeiro and want to share the news about your trip to South America with your friends. You will probably show them a few pictures and share some observations about the city, such as "it is huge," "it has beautiful views of the Atlantic Ocean," and "there are taxis everywhere." Through this act, you share some of the predications that constitute what Rio de Janeiro means to you with your friends.

But there is another possibility—that you will use one or more metaphors to capture what Rio means to you. For example, if you and your friends live in the American Midwest, you could say that Rio de Janeiro is "the tropical Chicago." In only three words you are thus able to evoke a series of predications about Chicago that are familiar to you and your friends and transpose them to Rio de Janeiro. You are inviting your friends to see Rio de Janeiro as if it were Chicago but in a tropical climate. With that, you also create a task for your audience. Which Chicago predications should be attributed to Rio De Janeiro? Your intent might be to highlight some of Chicago's most characteristic predications: it is a metropolis, it has a beautiful waterfront, it is full of social contradictions, and also that it has some dangerous areas.

Despite the possible ambiguities that this entails, this metaphor is still quite simple because it involves two members of a larger category—cities. Crossing categories can lead to much richer, and potentially even more ambiguous, metaphors. For example, a popular Brazilian song describes Rio de Janeiro as a "purgatory of beauty and chaos."[27] There is always an unexpected aspect to metaphors; in this case, the song asks listeners to see Rio as a purgatory, a place somewhere between heaven and hell, that combines aspects of heaven's beauty with hell's chaotic social tensions.

This metaphor is broad because it allows many interpretations; it conjures up a set of predications about purgatory that need to be somehow reconciled with the predications normally associated with an urban center. This is also a great example of how metaphors can produce meaning even when no single word in the lexicon captures anywhere near the possibilities offered by the tension between Rio and purgatory. It is simply impossible to cover all of the meanings of the metaphor using a predetermined set of concepts or explanations.[28] The metaphorical procedure of untangling the possibilities can effectively create a new meaning for Rio de Janeiro.

Like other innovative metaphors, "Rio is a purgatory" is an impertinent predication. It disrupts our interpretive expectations. When we interpret conventional predications, we operate in an almost automatic mode of interpretation–close to what Michel calls proto-interpretation. Thus, "Rio is a city" does not cause us amazement or cognitive discomfort. Conversely, an innovative metaphor is a shock to our semantic expectations because it makes a predication that is not generally used for that thing in our lifeworld. What do you mean by "Rio is a purgatory?" As a result, it becomes a puzzle that needs to be deciphered by exploring semantic networks related to Rio and to purgatory. The context, or framing, of a metaphorical utterance—in this case, the song's lyrics—gives us clues so that the impertinent predication can be reconciled with our semantic horizon through an interpretive process.

Finally, the tension between the vehicle (purgatory) and the tenor (Rio de Janeiro) makes the innovative character of metaphors explicit. This innovation does not end with a simple set of predications. It remains open and alive because, as we said above, "purgatory of beauty and chaos" is not just a set of words but evokes a comprehensive and complex semantic network. This complexity affords multiple entry points for interpretation and thus many different ways that it might be applied to Rio de Janeiro. One could, for example, start with how purgatory is linked to the punishment of the wrongdoings that destroyed paradise. Through such a lens, the social chaos of a violent metropolis like Rio can be seen as a kind of punishment for marring the natural beauty of the Brazilian coast with a large city. Productive metaphors like this one engender semantic innovation, bringing different concepts together in a relationship both blessed and cursed by tension and ambiguity. The clash of the impertinent predication thus fractures the conventional linguistic structures captured in dictionaries. By asking us to see something as something different, metaphors stretch our semantic horizons, making us see diverse aspects of our experiences in the world and creating new spaces of meaning.

Metaphorical Bundles

In praising metaphors, Aristotle says that the ability to create good metaphors depends on "an eye for resemblances":

> It is important to use aptly each of the features mentioned, including double nouns and loan words; but much the greatest asset is a capacity for metaphor. This alone cannot be acquired from another, and is a sign of natural gifts: because to use metaphor well is to discern similarities.[29]

Aristotle's choice of the verb *theorein* (to look at, watch, contemplate) is particularly relevant. It is related to both a visual physical capacity, for example, to see as a spectator of the Olympic Games or to look at a physical object. It also has a frequent use related to an intellectual capacity, the ability to contemplate theoretical truths. So, the etymology of the verb simultaneously points to a physical and mental dimension of the act of "seeing as." To metaphorize well is to develop a "sense of resemblance," a capacity to see one thing as another. In classical theory, the resemblance was one between words.

The Aristotelian interpretation became canonical in the establishment of metaphor as part of rhetoric in classical liberal arts. More recently, the contemporary semantic theory of metaphor, as exemplified by Max Black's work, moved the focus of this resemblance from the level of the word to the level of semantic predication.[30] Metaphors are no longer viewed as simple substitutions of words but are recognized as a more complex phenomenon, one that happens in the context of the sentence and represents a different kind of predication. The "seeing as" is no longer merely seeing one word through or as another; it is an exploration of the possibilities of the semantic field through novel predications expressed through metaphorical statements. So, instead of merely looking at two things and comparing them, one also looks around at the surrounding landscape and how they fit in. Paul Ricoeur calls this more contextual, sentence- level, perspective a tension theory as opposed to a word-level approach.[31] This new way of seeing allows for an exploration of how a new concept can be created out of the effort to reconcile the apparent paradox created by an "impertinent predication."[32]

Max Black suggests that the secondary subject of a metaphor, sometimes called the "vehicle," is better understood as a system rather than just as an individual thing.[33] For example, in "society is a sea," Black considers not only "sea" as an individual isolated object, but as a system of ideas, all linked to the central concept of "sea." He calls this semantic system the "implicative complex." It is the set of "associated implications" that could be attributed to the secondary subject.

This shift in focus from a secondary subject to a system of associations has a number of implications that will be important in our subsequent cognitive analysis and its application to digital metaphors. First, it opens a space for the perception of the various possibilities of a new attribution realized by the same metaphor. Taken as an open system of attributions, some of these attributes can be selected to reconcile the paradox afforded by the metaphor, and others cannot. Further, in principle, different sets of attributes could be selected, yielding different solutions to the paradox. The process of understanding a metaphor, therefore, is inherently about finding and selecting an appropriate set of predications. As we discussed above about "Rio is a purgatory of beauty and chaos," there may be no fixed set of predications related to the secondary subject that must be used or projected in the primary subject. The system of association highlights the openness of the metaphorical semantic attribution, its systemic nature, and the relative interpretative demands.

New Things in the World and in Language

When we need to "make sense" of new experiences in the world, we can leverage existing metaphorical bundles. Like new stars, metaphors are created in relation to other metaphors, forming their own constellations. These other metaphors push and pull the meanings of the new metaphor, and they create a region of overlapping predications. Lakoff and Johnson offer an extensive list of examples of these seminal metaphorical bundles, such as "argument is war" and "time is money."[34] So, once the basic metaphor "argument is war" is shared by a community, it is easy to create new meaning through other metaphors derived from that matrix, e.g., "he was attacking his position in the debate," or "her ideas offered a new line of attack to deprecated arguments."

As we briefly pointed out when exploring the *Back to the Future* example of motorized carriages, this mechanism of meaning expansion through metaphorical bundles is particularly useful for digital technologies aimed at creating new possible experiences. To make sense of these new experiences, designers need to link them to our linguistic and cognitive systems through metaphors. New technologies create things materially, and then metaphors make them communicable and understandable. In turn, these metaphors further create foundations, frameworks from which other related aspects of experience in the world can spring into the realm of language.

Let's take a look at the case of digital money, often called cryptocurrency. First, a mechanism was created so that the transfer of items of value could be done consistently and safely through digital artifacts. Once that mechanism was in place, it was necessary to tell people about it so that they might start using it. Such a mechanism is only useful once it can be understood by a large number of people who, in many cases, would not understand the technical details behind the mechanism. To this end, the metaphor of a digital currency, Bitcoin, was created. This metaphor tells the story that a set of ones and zeros encoded in a certain way and stored in magnetic devices is the same as currency. Or, more simply, in this system bits can be treated in the same way that we treat coins—we can buy stuff with them, give them to other people, etc. In this way Bitcoin creators have asked users to see bits—themselves based on a metaphor where we see the states of magnetic fields as representing numbers—as currency in a financial system. "Bits are coins" is the fundamental metaphor that in turn has been used to generate a metaphorical framework for a series of other metaphors. Once the initial metaphor is understood, the rest should be easily assimilated too, as they can be linked to the semantic network of "bits are coins." For example, we store real coins in a wallet, so it is easy to add the new concept of digital wallets. Similarly, to document the transfer of currencies between entities in the financial system, a ledger is used, so a digital ledger metaphor was created for the digital registration of transactions with Bitcoins.

For a technical enabler to become a technology that affects people's lives, it must be communicated; its meaning must be expressed through language. Metaphors are attractive for this communication because, as we have seen, they create new meanings from concepts that are already familiar to speakers of the language. Metaphors allow new technologies to be understood and viewed as familiar, even though they may be profoundly different from a physical perspective. The creation of technological enablers for transferring values through digital codes is only one step in digital technologies such as Bitcoin. For them to gain social momentum, they need to be communicated, accepted, understood, and trusted. And metaphors are a fundamental mechanism for doing this. The set of predications pertinent to this new technology is created as a metaphorical bundle derived from the originally impertinent predication: bits are coins. This seeing-as of a metaphor takes place over time, all the while transforming the set of meanings of people and societies as the metaphor becomes part of the linguistic lexicon. In the next section, we explore how metaphors are born, grow, and die in language through metaphorical cycles.

METAPHORIC CYCLE

Metaphors have a life cycle.[35] They start fresh, innovative, and defiant. They demand mental work to be solved because they represent a cognitive puzzle in the form of impertinent predications, e.g., "these bits are coins," and "this application is a wallet of coins." Obviously, there are problems to be solved when you hear such phrases for the first time. Even for those who know what bits are, never before did a sequence of bits mean the same thing as a coin does; after all, bits are abstractions and coins are concrete. Of course, many of us have interacted with a traditional bank over the internet. But even in that case, the monetary value was still represented in the currency approved and issued by the government, such as the US dollar in the United States. We understand that as if our money is literally sitting in a vault in a bank somewhere, even if the reality is somewhat more complex. To say that bits are coins is to create an impertinent predication. The solution to this conundrum can be simply to believe that the person who said that bits are coins was mistaken. Or, if the mistake does not seem obvious, you try to solve the problem by approaching the proper predications of currencies with those proper to a bitstream. You try to imagine how bits can be seen as coins. Which typical coin predications can be assigned to bits so that this new thing in the world makes sense? The proliferation of articles in the media devoted to explaining bitcoin suggests that resolving this particular metaphor is no simple task.

New metaphors invite us to think differently so that a new thing or experience or interpretation of the world can emerge in language. They show how language is open, expansive, alive. What we can communicate through language is not static, solidified in stagnant canonical lexicons; we can express and appreciate virtually any different aspects of our experiences. The world is open to new interpretations through new metaphors. We can find new meanings by choosing new interpretive paths to impertinent predications. These are key characteristics of live metaphors.

Over time, due to their frequent use, the predications suggested by new metaphors become part of the shared vocabulary of a social group. When it happens for a whole culture, the metaphor enters the dictionary or Wikipedia. They become ready-made bundles of predications that are commonly understood; they are no longer impertinent. They have been tamed. Tamed metaphors lose their openness and the rebel spirit that may have once challenged existing conceptual frameworks. They become stable and are solidified (some would say fossilized) into dictionary entries; they join the canonical conceptual framework of the language.

Poets want to nurture metaphors. They want to keep their wildness, their possibility of making us think differently, go out of the box, to explore new worlds, to boldly go where no man has gone before through impertinent predications. *For poets, metaphors are ends.*

Designers of new technology, by contrast, want to use the metaphorical mechanism to baptize the new things that they have created. They have a pragmatic goal: make people understand (and therefore use) their new technologies. So, they want to speed up the metaphorical life cycle. They want to kill metaphors as fast as possible so that the predications they want to convey become regular meanings in the world. *For people in tech, metaphors are means.*

To better understand how these metaphorical cycles happen, below we introduce the concept of the hermeneutic spiral, a concept that runs through the rest of the book. This idea is a model of how meanings evolve over time. It can be applied to any of our experiences in the world, but it is particularly salient for new technologies.

HERMENEUTIC SPIRAL

Consider the temporal schema below based on Paul Ricoeur's threefold mimesis that proposes that meaning is transformed in three phases organized in subsequent moments (t1, t2, and t3).[36]

t1: life-world A—As we extensively explored in the previous sections, at every moment of our lives we always have a set of meanings for things and experiences in the world—-e.g., justice, good, friendship, death, etc.—that shapes our interpretations, values, priorities, emotions, and actions. At the start of the transformation, we refer to this as "life-world A."

t2: semantic innovation—Innovative predications are proposed by new experiences such as conversations with friends, social media posts, books, works of art, and technologies. For example, John Doe is a bee, beauty is exemplified by Monet's *Two Sisters, courage is personified by Homer's Achilles, shopping is clicking buttons on a screen, and so forth. Many of these new predications are anchored by metaphors.*

t3: life-world B—As we interact with these new predications some of them become internalized in our life-world. As this happens, they reshape how we experience and value the world and ourselves. Such reshaping transforms our previous life-world, life-world A, into a new one, life-world B, through semantic innovations.

Perhaps a helpful initial example for thinking about the hermeneutic spiral is to consider a remarkable book that you have read or a memorable movie that you have watched. Works of art can be powerful because great ones offer a new way of looking at (a new horizon of meaning about) certain aspects of the world, our interpersonal relationships, and ourselves. Works of art can transform our life-worlds through semantic innovations. They propose new meanings for the concepts that organize many of our values and choices, such as love, death, deities, or work. It is not uncommon to hear people say that a particular book or movie "has changed their life."

Metaphors embedded in digital technologies are part of and drive a hermeneutic spiral as well. At first (t1), "to chat" had a specific meaning in Western societies. This means, as we saw above, that a set of predications were applicable to "to chat," such as "to chat is to have the face-to-face encounter." When (t2), the first message applications implemented the digital metaphor that a set of text typed on a computing device and sent to another user of the system is "to chat," they suggested an impertinent predication. At that point, "to chat" wasn't understood as meaning exchanging text messages over computers. It would be fair for someone to reject the idea and say, "but that's not what it means 'to chat.'" However, the adoption of messaging applications and the pervasiveness

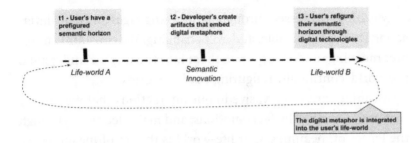

Figure 7. The Hermeneutic Spiral applied to semantic innovations implemented as digital metaphors.

of mobile devices inserted the digital metaphor into the semantic horizon of the people using those technologies, transforming the way they saw "to chat" and making it part of (t3), life-world B. Without even noticing, messaging apps also "changed their lives."

Once this cycle is completed, we return to (t1) because this is our new semantic horizon baseline. However, this process is not a circle, because we are clearly not back at exactly the same point (t1). What we could say, inspired by Ricoeur, is that we are at a new depth of the hermeneutic spiral which we could represent in our scheme as (t1-b). Most important for our purposes, through the hermeneutic spiral digital metaphors transform our horizon of meaning and therefore impact the ways we experience the world, our emotions, relationships, values, and actions. Such an approach to digital metaphors seems to us a more profound and productive way to assess the structural impact of digital technologies on contemporary societies once they reaches the cognitive and interpretive roots of the human experience of the world.

SPEAKING OF MEANING

Meaning is a fuzzy and intertwined set of predications that we attribute to things, people, and experiences. Such predications are dependent on the ways we use language to act in the world.

Meanings expressed through natural languages and daily interactions are open, flexible, and ever changing. We constantly interpret meanings and incorporate them into the way we understand the world, continually refiguring our life-worlds.

Meanings are so pervasive in our interactions that we can say that we see the world, feel emotions, and make decisions through the lenses of meanings. Our life-world is the set of meanings we use to make sense of our experiences.

Works of art, texts, actions, and technologies can all configure new meanings. Our life-worlds are refigured when we interact with these semantic innovations.

Metaphors are potent and compelling ways of conveying new meanings to new experiences and to things in the world by leveraging concepts that were already integrated into our life-worlds.

Digital metaphors are ways in which digital technologies are communicated to their users. By interacting with digital technologies, and by extension digital metaphors, users integrate such metaphors into their life-worlds, making them see, feel, and act differently.

CHAPTER SIX

COGNITIVE STRUCTURE, LEARNING, AND METAPHOR

In the previous, chapter we examined meaning through a philosophical lens, with a special focus on the idea of context, expressed in philosophical terms as predications. In this chapter, we ground these ideas in cognitive science, describing how people build models and understanding of the world by way of learning.

Our guiding principle is that learning is fundamentally about using experience to make better decisions in the future. More specifically, learning allows us to use the past to make predictions about the future. As Jeff Hawkins recently put it, "the neocortex learns a model of the world, and it makes predictions based on its model."[1] Thus, one important form of learning involves building models of the world, sometimes called "cognitive maps." We'll turn to another closely related outcome of learning, memory, in a later chapter.

Since our experience of the world is filtered by time, the most natural way to learn is to capture the sequences that we experience as we go through life. In terms of prediction, if you always experience C after experiencing A followed by B, then you ought to be ready for C to happen the next time you experience A and then B.

Thus, learning systems that are predictive should prioritize contiguity, things that are experienced close together in time. Indeed, this is one of the critical building blocks of learning. However, we do not learn and remember every sequence that we experience; our need for a prediction dictates that we synthesize the sequences we experience into something more abstract—patterns—such that they can be better applied to new situations. Because we are building a model of the world, such learning is normally statistical in nature. We can more easily remember things we do commonly, for example, and if we do things enough times, they even become habits. Much of learning, therefore, is slow and conservative, reflecting what statisticians would advise you about jumping to conclusions with too small a sample size. Thus, repetition is the other key building block of learning. As we shall see, one of the reasons that metaphor is so necessary and so powerful is that it allows us to circumvent this process and the need to rely on repetition. With metaphor we can find new meaning by leveraging new combinations of things we have already learned.

As was true with philosophy, our aim is not to introduce an entirely new model of cognition in this book, nor to survey existing models. Instead, as we drew from Paul Ricoeur for our philosophical foundations, we draw from a line of cognitive science that started with William James, but most importantly included Donald Hebb. Though James's work started in the nineteenth century and Hebb's work dates back more than seventy years, their most important ideas have held up and continue to influence cognitive science to this day. Before we get to their contributions, however, we contextualize why this perspective is important for our book.

EXPERTS AND METAPHORS

The story of our book begins with the public's first encounter with a new technology. Imagine a user, for example, browsing an app store looking for something new. That user will be presented with choices, more choices than they can easily navigate. For a new product to get chosen it must stand out; the user needs to know why they should choose this new thing instead of any of the thousands of other choices. For apps, this means that the app needs to clearly spell out what it is and why it should be selected. Many apps are extremely complex, consisting of hundreds of thousands of lines of computer code, and often require an extensive infrastructure to work. The information the user needs is more practical—what does it do and how do you do it? The user has a learning problem, and if it is not resolved quickly, they will simply move on to the next thing. As we have already seen, the answer to this problem is metaphor.

We must note that metaphors are not the only choice for effective communication. Comics are an increasingly popular alternative, including in technological domains. For example, Google hired cartoonist Scott McCloud and his colleagues to provide explainers for several complex topics such as machine learning.[2] Randall Munroe of xkcd fame has become renowned for his ability to explain things with comics, ultimately resulting in a book that specifically focuses on using comics to explain technology in simple terms.[3] Such comics can be expensive and time consuming to produce, however, and they also require having a cartoonist who fully grasps the technology themself. It is true that Google used a comic by McCloud to help introduce its Chrome browser.[4] But most developers do not have Google's resources, nor Google's stature with potential users. Meanwhile, even comics must rely on what McCloud has termed "structural metaphors."[5] Video represents yet another alternative, but the same arguments apply as with comics. Regardless, metaphors remain the primary method for communicating what a new technology is to the public.

The story does not stop with the metaphor; it merely begins there. From the point of view of cognition, it is a story about learning: how we build cognitive structure, how that process changes how we view the world, and how we can learn from others. With technology, that process almost always starts with a metaphor, and the choice of metaphor has myriad consequences, which we examine in detail in later chapters. This chapter is fundamentally about the process of building cognitive structure through learning and how metaphor can play a part in that process.

LEARNING THROUGH EXPERIENCE

Learning about the world is a singular challenge. In a nutshell, we have a small brain in a big world. The world is full of objects and actions, things are happening constantly, we are surrounded by information and by change. One challenge in learning about the world is that we are exposed to only a tiny fraction of it—and even that tiny fraction is rich with information, more than we can easily process. Meanwhile, our brain is limited in terms of the amount of information we can process through our senses and the speed at which we can process it. Yet, despite all of these challenges, we are able to identify useful patterns, make accurate predictions, and generally do well in the world.

We are born into a world of possibility. It is impossible to know what people will become important to us or what skills we'll need. A frog, by contrast, can come into the world with a brain wired and ready to catch flies. As a more adaptable species, we can only gain the knowledge we need through experience. All of which is to say that our success in life comes from our ability to learn no matter what kind of environment we might find ourselves in, whether an arid desert, a tropical rainforest, or a densely populated city.

The need to adapt to any situation is embodied by the basic unit of the brain, the neuron. Neurons are connected to thousands of other neurons, allowing them to be flexible and adaptive. The huge quantity of neural connections in the brain has led to the brain being referred to as the "connectome" in some circles.[6] The myriad neural connections that humans are born with represent possibility, the ability to mentally connect anything to anything. If the connections that we are born with represent possibility, then experience, by contrast, is a process of selection, strengthening some connections and weakening or even pruning others to better reflect the particular world that we live in. A Puluwat navigator who spends much of their life on boats in the Pacific may be born with the ability to learn how to thrive in an urban environment, but the life that they actually lead will emphasize a completely different set of knowledge and skills.

FINDING ORDER

The neural mechanism for the selection process in learning is known as "Hebb's rule," named in honor of Donald Hebb who first proposed it.[7] In truth, the principle that underlies Hebb's rule dates back at least to William James, who framed it in the nineteenth century in terms of "brain processes" instead of neurons.[8] Versions of it have been discussed by philosophers going even further back in time. Each version of such a learning rule is concerned with "coincidence detection," or times when two things happen at once. For Hebb, the two things are neurons firing, whereas for James learning was thinking about objects. In truth, James and Hebb's ideas are fully compatible and stated in very similar terms. Hebb's rule states that, when a neuron helps facilitate the firing of another neuron, the connection between them is strengthened. Thus, central to learning are the two concepts introduced in the previous section: contiguity and repetition. Contiguity, because learning takes place when two things happen closely in time, in this case two neurons firing. And repetition because neurons fire often. If two neurons fire together once, it indeed may be a coincidence. As it happens more, the concurrence goes from coincidence to pattern. The more the neurons fire together, the stronger their link will become.

James's version of learning states "when two elementary brain-processes have been active together or in immediate succession, one of them, on reoccurring, tends to propagate its excitement into the other."[9] It turns out that Hebb's rule, for reasons beyond the scope of this book, works at this higher level as well. When thinking of one thing naturally leads us to think of another thing; that process is called association. And thinking about two things at the same time is the essence of metaphor.

Thus, the basis for learning is association, and the mind is, as Kahneman put it in his seminal book *Thinking Fast, Thinking Slow,* an "associative machine." This is true at every level of cognition, from neurons to the objects of thought, which we refer to as categories or prototypes. It is also true whether thought is conscious or unconscious, or in what Kahneman calls "system 1" and "system 2." [10]

By itself, Hebb's rule has a number of implications for cognition. One of them is the reliance on repetition, which necessitates that learning new things is normally quite slow. This is a good thing. People who make inferences on too little evidence "jump to conclusions." In other contexts, this is known as superstition, e.g., deciding that a pair of socks is "lucky" because your favorite team won while you were wearing them. Coincidences happen constantly, so our standard for learning needs to be somewhat higher. What we need to capture in cognitive structure are patterns that can be exploited in predictions. Many readers will note that people do jump to conclusions all the time and that sometimes we can learn very quickly. So how do we square this with the idea that the basis of learning is slow? It turns out that the cognitive system can speed up learning under special circumstances, building on the basic mechanics provided by Hebb's rule. The emotional system, for example, can greatly accelerate learning in important situations (for some people, this would include their favorite team winning) by greatly intensifying neural activity. And, as we shall see, the language system, especially through metaphor, can also circumvent the normal process by taking advantage of existing cognitive structures.

Finally, it is important, and central to the thesis of this book, to note that learning is automatic and it is never "off." We are always learning, and every thought or imagining brings cognitive change. Learning is not something we can simply choose not to do, though clearly we can be intentional about it. Imagine a world in which learning required some cognitive process or decision-making. How would a young child manage to learn so much? Indeed, in their first few years children learn many of life's most challenging skills, such as language and walking. What if we missed something because we didn't mentally deem it important at the time? Learning is too critical to be anything other than automatic and ubiquitous. It is true that we can consciously choose to intervene—we might repeat something to ourselves to generate extra repetition, or we might focus our attention somewhere else—but Hebb's rule is always operating.

Emotions and Learning

The relative slowness of learning is fine for the most part. Coincidence detection and association are statistical operations, and statistical operations need largish samples to be truly effective. But the world does not always support such sluggish learning. Sometimes we really do need to learn something based upon a single event, e.g., if your ancestors were walking by a cave and encountered a dangerous animal and barely escaped with their lives, they would have been more likely to live longer if they strongly associated the cave with danger. This is where the emotional system can help. Arousal in particular allows the cognitive system to boost learning, essentially short-circuiting the normal statistical nature of learning, allowing for a kind of "burn-in" of important events.

It is important to note that this does not require an additional learning rule. Hebb's rule, after all, is predicated upon correlated neural activity. Arousal focuses and intensifies neural activity in the brain through the release of the neurotransmitter dopamine, automatically creating more correlated activity by ramping up internal repetition.[11] Abstractly, we can think of arousal as a kind of measure of importance. Some things arouse us because we are hardwired to respond to them by evolution—pleasure, pain, animals, loud noises, sex, violence, etc. Other things arouse us because we have learned that they are important—the presence of our boss, a final exam, etc. All of these translate into a heightened state of arousal, which in turn releases dopamine in the brain, which has the effect of amplifying learning. Therefore, we can crudely think of arousal as a kind of dial that turns the speed of learning up and down. Thus, some highly emotional events—the assassination of a national leader, a terrorist attack, a near-death experience—are essentially unforgettable.[12] Less intense variations are also commonly seen, especially given the relationship of pleasure and pain to arousal.

Hebb's rule has another important implication with regard to emotions. When we think about important events after they occur, the nature of association ensures that we automatically feel the emotions, and by extension the pleasure and pain, that happened during the event. Thus, thinking about an encounter with a dangerous animal is likely to bring back the feelings of fear that occurred during the encounter. This helps us to determine the right course of action. In his seminal book about emotions, *Descartes' Error*, Antonio Damasio recounts the case of a man, whom he calls "Elliot," who doesn't associate emotions with events due to brain damage. Damasio shows in careful detail how debilitating this is, as the man is virtually incapable of distinguishing good decisions from bad:

> At the end of one session, after which he had produced an abundant quantity of options for action, all of which were valid and implementable, Elliot smiled, apparently satisfied with his rich imagination, but added: "And after all this, I still wouldn't know what to do!" [13]

Negative consequences of this associative link, on the other hand, include problems like post-traumatic stress disorder (PTSD).

The associative nature of emotions is important in technology because tech developers are able to take advantage of these linkages when choosing how to present their technology. Social networks, for example, are framed in terms with positive emotional affect, such as "friend" and "like." As Sismondo has noted, metaphors do not simply change the target of the metaphor, but both the target and the source: "To say that the world is a machine is not just to shape our conception of the world, but also, eventually, to reshape our conceptions of machines; we make the world appear a little more machinelike and machines a little more worldlike."[14] In other words, when Facebook tells us that our connections are "friends," it not only helps us to understand this new category by pulling relevant features from our existing "friend" category, but it also will work to change the "friend" category through its association to the Facebook construct.

The choice of positive metaphors, such as "friend" and "like," is effectively free goodwill for an app's designers thanks to association. In psychology, this effect is known as metaphoric transfer.[15] With metaphoric transfer, changes in how someone feels about the source of a metaphor impact on how they feel about its target. Williams and Bargh did a pair of studies that show how this works and provide a kind of blueprint for how it could be exploited. In one, they took advantage of the common metaphor that links distance to emotional attachment.[16] They asked participants to place two dots either close together in space or far apart. Participants who placed the dots farther apart subsequently reported a weaker emotional bond with their family than participants who placed the dots close together. In the second experiment, they looked at similar effects with warmth metaphors.[17] In this case, simply holding a warmer or colder beverage in their hands made participants describe a target as having a warmer or colder personality.

Clarity, Confusion, and Boredom: Seeking or Avoiding New Experiences

There is another side of the emotional coin that is less well studied, related to our need for information. As a species, we want to understand our world. Our success has not come due to size, sharp teeth, or claws; it has come from our ability to acquire and use information. Intrinsically, this means that we tend to seek out information, but we must also balance this desire with our need for safety. In machine learning, this is commonly framed as a trade-off between exploration and exploitation. Exploitation allows a learning system to reap the benefits of experience, but exploration may lead to the discovery of new and better paths and methods. Systems that exploit too much may never find the best available methods, but systems that explore too much risk getting into dangerous situations, or find themselves wasting time and resources when easy success might have been possible.

In cognition, Ivancich,[18] building on work by Kaplan,[19] has framed this trade-off in terms of motivational pushes and pulls. Boredom, for example, pushes us towards exploration and trying new things. When we have our environment figured out, it is a good time to explore a little. Confusion, on the other hand, is at the other end of the informational spectrum; it pushes us to retreat back to what we know and understand. It is dangerous, after all, to be confused, as it signals that we are unlikely to make good predictions and decisions. Further, as any school kid knows, when presented with a confusing new topic, confusion can be cognitively painful. What Kaplan and Ivancich call clarity refers to a range of related experiences, including "interest, fascination, mystery, intrigue, the thrill of discovery."[20] All of these things pull a person deeper into an experience, with the clarity providing pleasure that can range from mildly pleasant to intensely joyful. And all these forms of clarity invite exploration with the promise of new frontiers of knowledge.

All three of these mental states play key roles in the world of apps and technology. Bored with the same old apps that you have been using forever? App stores have an ever-growing supply of new ones to keep you engaged. But if a new app is confusing in purpose or usage, it is easy to delete it in favor of something else, making the ability to present them in simple terms crucial to their success. Meanwhile, apps that bring any of the aspects of clarity can pull new users in with the promise of pleasures to come and continue to engage existing users with new things to explore.

Attention

In our discussion of Hebb's rule, we noted that it is always on and therefore that learning is largely automatic. We further noted, however, that there are times when we want to exert some cognitive control over this process. This is where attention comes in.

As we have explained, we humans have a small brain in a big world, and the amount of information available to us greatly outpaces our ability to process it. This implies that we have limitations on what we can process at any given time. Evidence of such limits can be seen in the small number of items that we can hold in our head at a time.[21] Before we dive into the mechanics of attention, it is worth connecting this idea, that people have a limited capacity to process information, back to the philosophical treatment in the last chapter. For extensive concepts (like "Rio" for a native Brazilian), it is impossible to hold every aspect of the concept in one's head at once. Thus, context must help determine just what aspects that we attend to. In terms of what we are processing, there will be a significant difference in hearing "the home of the statue of Christ the Redeemer" and "the host city of the 2016 Olympics." Returning to association, in the first case it is likely that thoughts of art and religion are primed, whereas in the other it is more likely that we will prime thoughts of sports and international events.

It is common to think of attention as a unitary concept, but

William James, sometimes called the father of modern psychology, noted that there are in fact two kinds of attention.[22] The first, "directed attention," requires effort; we even refer to the cost of the effort as "paying" attention. James contrasted this with "effortless attention" or "involuntary attention," when material is so compelling that it is difficult to ignore. The involuntary quality of this attention stems from our evolutionary history, when there were things that we needed to pay attention to in order to stay safe. As Kahneman puts it, "orienting and responding quickly to the gravest threats or most promising opportunities improved the chance of survival."[23] Movement, for example, might signify the presence of a dangerous animal, so there is value in attending to it.

In the modern world, the two types of attention are often at odds in a way that pits our intention against our inclination. Consider a student trying to study a difficult subject in their dorm room while their roommate is watching an action movie on television. The action movie is full of the sorts of things that trigger involuntary attention—loud noises, violence, color, motion, etc.—and, as the name suggests, their pull on the student's attention is involuntary. Thus, the student is inclined to watch the movie, which is at odds with their intention to study. The student who wants to study has a choice in how they overcome this distraction. They can leave or they can attempt to deploy directed attention, buckling down to focus on their task. For environments such as the one the student finds themselves in, this is difficult, as directed attention is sapped the more it is used. As we shall see later in this book, many apps in the digital world are optimized to capture our involuntary attention, in turn putting a severe strain on our directed attention.

Stephen Kaplan and his colleagues, among others, have been studying the costs of directed attention fatigue for decades.[24] They have found that directed attention is a limited resource, and that the consequences of overusing this resource can be fairly dire, impacting cognitive performance on many tasks, and even health.

Roy Baumeister and his colleagues have done similar work and term the phenomenon "ego depletion" having noticed that when you force yourself to do something you are less ready to take on the next challenge.[25] Meanwhile, the benefits of experiences that are restorative to directed attention, for example, walking in nature, are many. For example, an early study done by Cimprich on patients during treatment and recovery for breast cancer showed that patients in an experimental group instructed to spend time in nature, such as walking or gardening, saw significant improvement in cognitive tests and in quality of life metrics over patients in a control group, and on average were more likely to return to work and got back to work sooner than the control group.[26]

Students in the midst of finals at the end of a long semester and workers regularly putting in long hours are generally very familiar with the impact of directed attention fatigue. Patience, for example, relies on directed attention, so when directed attention is in short supply, so too is our patience.

Involuntary attention is important to our story in several ways. For one, it is commonly deployed by digital technologies to force us to pay attention to them. Think of the noises, vibrations, and other tricks used by notification systems. They are designed specifically to wrest control of our attention. And we willingly pay this cost, because we are afraid that otherwise we might miss out on something important. Meanwhile, digital apps are the source of a nearly infinite supply of distractions. There is always another funny video to watch on TikTok, or a new post to read on Twitter. In a world where people increasingly spend their time online, this can severely stress our attentional capacity. In turn, the constant competition between apps and services for our attention, the so-called attention economy, has resulted in a kind of arms race where the individual actors are better and better at getting and holding our attention.[27] Metaphors of the type we are discussing are the primary tool to enable users to connect with a potential audience.

Digital metaphors are designed to explain and engage. All of this presents a constant stress on our directed attention. In turn, people whose directed attention is fatigued will make worse decisions, be more irritable, and have worse health outcomes. In and of itself, these are bad enough outcomes, but a simple decision for a cognitively fatigued person engaged in watching TikTok videos is to keep watching. It might seem that letting involuntary attention win is a way to help directed attention recover, but Kaplan's ART (Attention Restoration Theory) model further divides involuntary attention into things that provide "soft" fascination, such as nature, and those that require "hard" fascination and occupy the mind fully.[28] When in nature, our mind is free to wander and thus can recover. By contrast, when we are engaged with an app, there is little opportunity for contemplation. Kaplan's group has found that activities that provide soft fascination are far more restorative than those that require hard fascination. In short, our constant engagement with the digital world comes at a high cognitive price.

Scaling Learning

The learning system that we have described so far works well for experiential learning—the world around us. Hebb's rule serves to emphasize patterns in our experience, and through emotions and attention the learning system is able to further refine what we learn. Attention is the primary way we deal with the problem of the volume of information that we take in. However, we are still limited to our own experience, which represents but a tiny sample of the world. To return to an earlier example, if one family member walked by a dangerous location, it would be a shame if the only way for other family members to learn of the danger was for all of them to walk past it themselves. To learn beyond our own experience, we need to be able to go beyond our experience; for humans, that ability is provided by language.

EXPERIENCING THROUGH OTHERS' EYES

Language allows us to experience things without living them. When another person says to us "The robot kicked the soccer ball," we can imagine the scene in our own heads. The major difference between seeing a robot kick a soccer ball and being told about it is visual detail. Ultimately, in each case a representation of a soccer ball will become active and a representation of a robot will become active, as will a representation of the robot kicking the ball. The details of what those representations are like are the subject of long debate in cognitive science, which we do not explore here. Instead, we will merely point out that the two experiences will show significant overlap in terms of which parts of the brain are active.[29] Harkening back to Hebb's rule, this means that we can learn in very similar ways through language as we do through experience.

This is amazingly useful of course, as it greatly widens our ability to learn about the world around us. Once we could learn from others through language, we could greatly speed up the learning process and find out about parts of the world that were not immediately accessible. However, it also meant that we might learn things that are not actually true. We can "experience" things that have not actually happened. Sometimes this is benign, as with a good novel, and sometimes it is actively harmful, as with disinformation campaigns.

Normal descriptive discourse is a useful way to tell someone else about something that happened, but sometimes it isn't enough in terms of what we want to communicate. Sometimes we want to tell someone about something new. This sort of communication cannot simply rely on activating internal structures already present in the listener. Instead, it requires building those structures from scratch. This presents a real problem because, as we have already seen, normal learning, and especially learning new things, can be very slow and can rely on lots of repetition. This relative slowness may be adequate when reading a book or listening to a lecture, but in the course of normal conversation, it does not cut it.

Metaphor is the solution to this thorny problem. Metaphors allow us to use things that we already know about to create new things that we did not know about previously. Thus, the Shakespearean metaphor "time is a beggar" can communicate a new idea by drawing upon pieces of what we know about beggars and combining them with pieces of what we know about time. Another Shakespearean metaphor "Juliet is the sun" provides a quick description of Juliet, again based on the idea that certain features of the sun can be shared by people. Not only is the metaphor poetic, but it is much more compact than simply listing Juliet's relevant features.

Metaphors are useful precisely because they allow us to leverage things that we already know about to quickly restructure existing categories or to learn brand-new ones. We can view metaphors, therefore, along a continuum, At one end, they are used in a mainly descriptive way, describing things that we already know about, but highlighting salient features or presenting the thing in a new light. At the other end, metaphors can be used to create brand new categories. To explore this continuum, we present three scenarios at different points along the continuum demonstrating three important uses of metaphor. The third of these scenarios represents the one that we are most interested in in this book.

Three Types of Metaphor

The first scenario (Figure 8 (a)) is probably the most common in everyday usage. In this type of metaphor, the speaker uses one category to emphasize certain aspects of another category. A typical example would involve two people discussing a mutual friend. One of the friends could metaphorically say "the man is a wolf." As both friends know the man in question, the goal of the statement is not to create a new category, but to highlight things about the category that they already have. In this case "wolf" will associatively prime certain aspects of the category at the expense of others. There are many possible features that might come to mind when talking about a man, but the context "wolf" will shape which ones are active during the conversation. Thus, a different set of features might be active than would usually be the case when talking about a person. Hebb's rule will work to further strengthen the link to those features, as well as connect the representation of the man more closely to the representation of a wolf. This is emblematic of the fact that our mental representations are never truly static, thanks to Hebb's rule. It is important to point out here that this is a case where Hebb's rule can occasionally be insidious. The listener might not actually agree that the man is wolf-like, but without some mental effort the associative linkage will be strengthened regardless. If you are told something is true often enough, it is difficult to avoid having it creep into your mental structure.

In the second scenario (Figure 8 (b)), one person tells someone else about something new. In this scenario, however, the new thing isn't completely novel; it is just a novel example of an existing category. Here, we could have two friends discussing a third person, but one of the friends does not know the third person. So, when the other friend says "the man is a wolf," the sentence has a very different effect than that of the first example. The other person now knows two things about the man being described: that he is a man, and that he has some things in common with wolves. That person knows a lot about people and also about wolves, and they can use this knowledge to build a quick framework for the man using the features that the two things have in common—probably the same set of features as in the first example. In this case the result of the metaphor is a brand-new category made by combining two things that are already known. Further, the new category is a subclass of one of the original categories, in this case, people. Returning to our theme that learning is about prediction, in a case like this, with just a single sentence we have enough information about this person that we have never met to be able to predict quite a few things about how they might behave. This is one of the major benefits of metaphors; they can transmit a tremendous amount of information with only a few words.

The third scenario (Figure 8 (c)) is rarer in normal conversation, but is important for metaphors related to technology. In this scenario one person tries to describe something completely novel to another person. This new thing isn't simply a new version of something the second person already knows about, but is fundamentally different in some important way. One scenario might involve someone trying to describe their scientific breakthrough to a friend who is a complete novice in the field. The person would have to describe the breakthrough in terms that are familiar to their friend. This would be similar to what we saw in the example at the start of this chapter, and metaphor is the normal way to do this, as occurs with many new technologies. Thus, a device made of complex circuitry, wires, and the like can be described as a way to have a conversation at a distance. What is important about this scenario is not only that a new category created, but also that the existing one is modified. In the second scenario, describing someone as a wolf doesn't change how we think about people generally, just about the new person. Indeed, we probably know other people who could be described as wolves. In this new scenario, however, the metaphor represents a violation of the original category. In the world before telephones, conversations were not held at a distance; closeness was an essential feature of what it meant to have a conversation. It is such a category violation that makes the technology useful in the first place. Thus, the metaphor creates a kind of schism in the category and generates two new subcategories—e.g., conversations close together and conversations at a distance. The thing that differentiates the two subcategories is the new device. With other metaphors, sometimes called "novel" metaphors, the resolution is not so straightforward, because there might not be any features that appear to be in common.

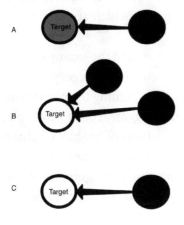

Figure 8. Three types of metaphor. Each oval represents the representation of a category involved in a metaphor. The fill of each oval represents how well known the concept is to the listener. The arrows represent structural influence due to the metaphor. (A) In this case both the metaphor's source and target are known to the listener. The metaphor suggests a particular way of thinking about the target and will mainly strengthen selected internal associations. (B) Here the target of the metaphor is new. E.g. a person that the listener hasn't met. In this case in addition to the source an additional influence on the new category will be the class that the new thing is an example of. E.g., a person. The new object will be constructed from a combination of the target and the known class and will be a new example of the known class. (C) In this case the target is completely unknown and also isn't simply an example of something that is known. In such cases the new class can only be influenced by the source of the metaphor.

A Second Metaphorical Dimension: Novelty

It is fairly simple to see how association is related to metaphor, e.g., there are certain features of wolves that are often features of men. Thus the "the man is a wolf" metaphor should associatively activate those features. Indeed, one leading cognitive model of metaphors is based on this idea.[30] This is not, strictly speaking, the whole story. It works well for what we call category 1 metaphors, where the metaphor concerns two well-known categories, but it seems to fail for more complex metaphors, which are sometimes termed "novel" metaphors.

In discussing novel metaphors Ricoeur used the example "time is a beggar" from Shakespeare. This is clearly a category 1 metaphor, because the vast majority of people watching the play would have solid categories for both time and beggars. For most people, however, the features of time and the features of beggars will have little to no overlap. Indeed, when we attempted to resolve the metaphor ourselves, we were unable to come up with a good set of features. The primary meaning of "beggar" involves someone asking for food or money, which does not seem connected to time, and indeed seems to be at odds with the common metaphor "time is money." The intersection of closely related features seems to be empty. To understand the metaphor requires the context provided by the original text; it cannot be a simple matter of word substitution. In Shakespeare's play *Troilus and Cressida,* the character Ulysses says:

> Time hath, my lord, a wallet at his back,
> Wherein he puts alms for oblivion,
> A great-sized monster of ingratitudes:
> Those scraps are good deeds past; which are devour'd
> As fast as they are made, forgot as soon
> As done. [31]

Shakespeare's language guides us through our associations, providing context to set up (or prime, in the language of association) what comes next. Here we can see that the metaphor requires either a very particular view of beggars, that they are "monsters of ingratitude," or at least that such a view fits a particular class of beggars. This particular metaphor is challenging, because such a view of beggars may not typically be part of the reader's mental model of beggars (their sociocultural context), and it is also revealing because it yields insights into the speaker's beliefs.

Shakespeare handles this complex metaphor elegantly. The primary subject in this case, time, is abstract, and thus not apparently easy to unite with a category as concrete as "beggar." Shakespeare begins by personifying time. From there, Shakespeare alternates between aspects of the beggar, using terms like "alms" and "scraps," and aspects of time, with allusions to oblivion, the past, and forgetting. Thus, Shakespeare is doing the work that we would normally do ourselves when resolving a metaphor in terms of determining the relevant features. During this process, Shakespeare creates a vivid new object for the listener that contains aspects of each category. This new object is only possible because of the system of associations the listener already has in place.

As we argued in the previous chapter, technology developers also rely on metaphors to create new things But they cannot afford to do so with metaphors as complex as this one; they rely on the associations being crystal clear to everyone. This is not to say that Shakespeare did not engage in such metaphors himself; indeed, one of his most famous metaphors and one we have seen before, "Juliet is the sun," is very simple to decode with little or no additional context.

The difference in how people process simple versus novel metaphors has been well studied in cognitive science and can be seen, for example, in MRI studies of the brain. Subjects who are decoding complex metaphors actually need to recruit additional areas of the brain to make sense of them.[32] There are a number of theories about how this works, but it is clear that more is going on than simple priming.[33] A discussion of this is beyond the scope of this book, but it is important to note several things about such cases. First, "solving" such metaphors is pleasurable. The cognitive origins of this pleasure are likely tied to the idea of clarity that we discussed previously. To solve a metaphor is to understand, and to understand is to know that you are likely to respond effectively. By contrast, to struggle to understand is to be confused, which is painful. Thus, our second point: that not being able to see the resolution of a metaphor is unpleasant. Finally, as we saw with "time is a beggar," the resolution of a complex metaphor may be strongly tied to personal experience and cultural norms.

Metaphor and Creativity

So far, we have framed metaphor primarily as a vehicle for communication and learning. Once the cognitive system had the capacity to create and use metaphors, it became possible to use it for other purposes as well. One such purpose, and one closely aligned with the idea of communicating novel concepts through metaphor, is using metaphor as a fully creative act. Albert Einstein's famous thought experiments, or *Gedankenexperiments*, provide a good example of the intertwining of the two purposes. Einstein often referred to himself as a visual thinker and described a number of his breakthroughs in those terms. For example, when he was young and interested in the properties of light, he would imagine himself running alongside a beam of light. He later cited this as inspiration for his work on general relativity. In turn these thought experiments are also useful ways of explaining his ideas.

That metaphor can be used as a creative act is a natural extension of the major purpose of cognition, which is to make predictions. Returning to our example of seeing a dangerous animal near a cave, we can imagine what might happen if we were to return to the cave and predict that we could have a bad encounter with the animal. In what is perhaps the most famous quote in cognitive science history, Kenneth Craik put it this way:

> If the organism carries a 'small-scale model' of external reality and of its own possible actions in its head, it is able to try out various alternatives, conclude which is the best of them, react to future

situations before they arise, utilize the knowledge of past events in dealing with the present and future, and in every way to react in a much fuller, safer, and more competent manner to the emergencies which face it.[34]

In other words, our mental models allow us to run experiments in our head in a fashion analogous to Einstein's *Gedankenexperiments*. It is much safer to run a mental experiment about a possible return to the cave than to find out by actually going. With mental models we can safely "play" in our heads by changing various aspects of the model or its input, e.g., you might think "what if I went back to the cave, but this time with a group of friends." Metaphors extend that capability by allowing us to connect two things that we may not have connected previously. Put another way, our mental structures are models, and metaphors provide a mechanism to change the inputs to those models. Once we have the mental structure of a wolf, for example, we can plug a man into it. The amazing thing about our cognitive system is that it is able to ignore the contradictory parts that this almost necessarily fosters, and instead focus just on the parts that fit.

Advertisers and app developers use these abilities to get people to try their products, often painting a mental picture of how their products will be used and the ways that it will make customers happy.

IMPLICATIONS FOR USING AND ACQUIRING MEANINGS

Why Are Metaphors So Powerful and So Ubiquitous?

Hebb's rule has many important implications for the topics of this book. First and foremost is that learning is always "on." We don't choose to learn, and we have little direct control over it. Imagine a world in which we did have control over our learning. How would children learn the right things? What if we made mistakes in what we chose? Learning is too important for organisms such as humans for it to be anything other than automatic. Second, the statistical nature of learning requires that it normally be quite slow. Jumping to conclusions with little experience or knowledge can be fatal in a dangerous world. And yet, it seems as if learning new things is often easy. That is the power of metaphor.

Here we remind the reader that we use the term "metaphor" loosely, to describe a large number of linguistic techniques that have the common feature of referring to something as something that it is not.[35] The power of metaphor is that it allows us to build new concepts, not by experiencing them again and again, but by leveraging our experience with other concepts that we have already learned. In describing Juliet, Romeo need not laboriously list all of her features and hope that we can associate them with her; instead, he can simply describe her as the sun, a concept that we all know and one that allows us to easily infer her features. Facebook could have chosen a new name for one of its features, then spent time trying to teach us what it was and why it was useful. But it was simpler to call it "friend" and gain all of the positive implications of that word. Friends are people we like, want to catch up with, etc. So, metaphors are cognitive shortcuts, allowing us to bypass the usual rules for learning and to take advantage of the learning that we have previously done. These shortcuts are so useful that they represent a significant percentage of our communication, about 20 percent, according to one estimate.[36]

Further, metaphors can leverage other aspects of cognition that also aid learning, especially emotion. Friends are generally people who bring us pleasure, and thus friendship is an inherently pleasurable concept. Therefore, metaphors that build on friendship, or love, or any other concept strongly associated with emotions, automatically gain those associations. The idea of connecting to friends is pleasurable because in our own past connecting with friends has been pleasurable. Facebook and other social networks not only get a free association with this pleasure, but the pleasure actually can boost the learning process.

Of course, shortcuts have many advantages in terms of efficiency, but they can also have drawbacks. Facebook friendship isn't really the same as friendship in real life, after all. Metaphors have potential drawbacks as learning shortcuts as well. A successful metaphor relies on its recipient having the right model, for example, in "time is a beggar," having a model that beggars are, or at least can be, "monsters of ingratitude." When someone describes another person as a wolf, they probably do not mean that they howl at the moon at night or live in a cave. Thus, it is not enough to merely create a metaphor. It is important that the right set of features is extracted. Even if the metaphor is properly understood, the metaphor may not actually be appropriate, which may lead to the wrong conclusions. Even common metaphors can have this problem. For example, metaphors to describe the stock market as "climbing higher" or "falling off a cliff" impact how people react to financial news.[37] In the first, stocks are described as if they were agents with the implication that they have purpose and will. Those extra consequences, unintended as they may be, come along as a result of association, with the result that people exposed to the metaphor tend to feel that those climbing stocks will continue to do so tomorrow. By contrast, an object metaphor such as "falling off a cliff" does not imply such agency and so does not generate any further expectations. Of course, what is a problem in one domain may be a valuable feature in another. Returning to Facebook, the implicative complex surrounding friendship is very positive, and thus the associations of those positive feelings will tend to be reflected back on Facebook due to the power of association as expressed through metaphoric transfer.[38]

As we have seen, metaphors are so useful because they take advantage of cognitive structure, in the form of associations, that already exists. This is also true of the emotions associated with such structures. The "like" metaphor is an example of how social networks can use these associations to drive user behavior. On apps like Facebook, liking something is a positive act that in turn sends a signal back to the posting person that their post is appreciated. This sets up a feedback loop where people are eager to gain more likes, and the positive emotions that accompany them, and thus are likely to post more often. But this loop also creates expectations that their friends will like everything they post—if our friends have liked our posts before, our learned model will expect them to like them in the future. This situation benefits social networks, because it drives engagement and encourages more posting, in turn creating more content for the network. We come back to this example in more detail in the final part of this book.

For a person or group developing a new technology and desiring to create new meaning, there is a choice. To get users to go through the contiguity and repetition necessary to build fresh cognitive structure would be slow and tedious, which is not a situation conducive to getting buy-in to a product. Emotions offer an alternative, but not one that is realistically scalable (the YouTube phenomenon of unboxing videos does show that this approach works for some people). It is not surprising then, that time after time, product after product, developers choose to rely on metaphor.

The Cognitive-Hermeneutic Cycle Revisited

Many people were first exposed to the poet Amanda Gorman during the US presidential inauguration in 2021. For those people a number of features may have stood out. First and foremost was the poem she recited, which was widely hailed and has since been turned into a best-selling book. Then there was her clothing, which also caused a stir and was praised for bringing brightness and color to a stage where participants often dress in drab grays and blacks. Then there was her race, which also stood out amid a largely white group on stage, and her youth, which was similarly notable. Some combination of these things would be associated with the nascent representation of her in people's mind, knitted together by Hebb's rule and forming an initial set of predications.

Subsequent news stories would have solidified some of these predications and introduced new ones—she is mentored by Oprah, she had a speech impediment growing up, etc. Each of these encounters in turn serves to strengthen some of the existing connections emphasized in the articles, again through Hebb's rule, and to build new ones, thus making the cognitive structure associated with her deeper and richer. Later, a series of articles appeared describing the difficulties involved in translating her work into other languages and how race complicates that work.[39] On social media, Gorman announced that even after her speech she had been profiled by law enforcement.[40] Anyone reading such articles would add depth and nuance to their understanding of her and her work. Thus Hebb's rule, and the fact that any encounter with a complex figure will only reveal part of what is going on, ensures that meaning is always in flux, gaining new components and emphasizing others with each encounter. Each of these encounters is one arc in the cycle (Figure 9). In the meantime, as in this example, the world changes between our encounters. In this case, Amanda Gorman has a social media presence, has released other work, and has meaningfully reacted to her own experience of being on that stage. These actions, in turn, have been picked up by the media and reported out to the world. That same media pays attention to how many people click on stories about her and uses that information to decide whether to write more. Other predications involved, such as the role of race in US politics, also continue to evolve in parallel.

This continuous cycle is also in play with metaphors and technology, and it can even be magnified by the connections between users and developers. First, the metaphor and the artifact are linked in our mind during our initial encounter with the technology. Then, each subsequent encounter will strengthen that metaphorical connection and provide an opportunity for further learning and depth. In the meantime, every individual's experience with the technology will be different. Each will come to their encounter with a different set of life experiences and cognitive models. In

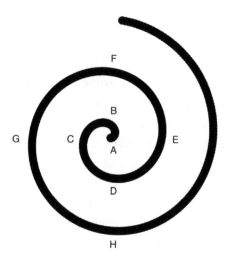

Figure 9. A simplified view of an example spiral. Starting at A: Gorman reads her poem at the inaugural. B: The audience watches and reacts. C: The media also watches and reacts producing new stories about her including, D, interviewing her and then E makes the stories available. F: The audience consumes these stories, learning more about her. G: the media notices the strong reaction to her poem, page views of stories, etc. H: she posts on social media, releases a new book, etc.

turn, they will use the artifact differently and in different contexts. Inevitably some of them will learn new, unanticipated, ways of interacting with it, or will at least find things about it that they wish to have changed. Meanwhile, the technology community has created efficient mechanisms for this feedback to be channeled back to developers. In turn, developers can use this information to iterate on the next version of the technology and the cycle will be renewed.

SPEAKING OF COGNITIVE STRUCTURE

Cognition is centered around making predictions. This places a premium on finding recurring patterns and sequences. The primary mechanism for learning such patterns is known as Hebb's rule.

Cognitive structure is associative. Thinking of one thing that we frequently experience in the presence of another thing will naturally make us likely to think of the second thing. This also applies to emotions. In turn, emotions can help speed up learning.

Attention is crucial to choosing what cognition will focus on next. Attention can be thought of as two systems, an involuntary system that automatically responds to intrinsically important stimuli, and a voluntary and directed system which can be consciously deployed. The directed system requires effort and will eventually fatigue.

Language provides a way of scaling learning to include things that we do not directly experience. Metaphor can accelerate learning by taking advantage of existing cognitive structure. Metaphor can either reconfigure existing categories or quickly build new ones.

PART III

DIGITAL METAPHORS IN THE WORLD

Next, we turn our attention to some of the major digital metaphors in today's world. Our thesis is that these metaphors are increasingly mediating nearly everything we do. As such, we begin with a chapter that explores this idea. We then follow with a series of chapters examining different examples of such mediation. As we have noted, digital metaphors exist in a hierarchy, and thus our collection of chapters begins with a foundational metaphor: touching is selecting. This is an example of a metaphor so natural that it is easy to forget both the fact that it is a metaphor at all, and the complex software necessary to make it work. Next, we look at social media and how metaphors like Facebook's use of friendship are reshaping how we think about relationships, and in turn how social media companies are in a competition for our attention. We then revisit our example of texting and conversation and examine the evolution of metaphors relating to texting technology. In this chapter in particular, we see how the cognitive-hermeneutic cycle can impact not only users but the technology itself, as texting began with clumsy metaphors around telegrams and mail and evolved to more straightforward and effective metaphors of chatting and conversation. The relationship of videos and photos to memories are explored in the subsequent chapter. In this chapter, we see that not only are tech companies increasingly using memory as a metaphor for their products, but also that these products can actually change the very memories that they are representing. We conclude this section with a look at smartphones. In this case, the initial metaphor—a phone not tethered to a single location—has become completely inadequate for how we now use these devices. In the chapter, we explore more appropriate ways of conceptualizing these devices, centering on ideas of identity and their ubiquitous nature.

CHAPTER SEVEN

MEDIATING THE WORLD

The framework we are proposing allows us to address a fundamental question about technology: Are we shaped by technology, or do we shape it to suit our needs? This question is typically phrased to suggest that the answer is one or the other but, as we have shown, the language chosen to frame an issue can change how we think about that topic. In this case, there is a continuum marked by two extreme views of a technology's agency. At one end of this continuum is technological determinism, which emphasizes how digital technologies define our personal and social habits in an almost autonomous way. In this theory, the agency of the relationship between users and technology lies completely with the technology. Users are portrayed as passive, easily seduced by the coolest tech widgets on the market. Proponents of this vision emphasize how, for example, smartphones determine how we communicate and disrupt traditional forms of conversation and social engagement in ways that we have no control over. In this view, the power of digital technologies is such that the mechanisms by which they are created and the ways users integrate them in their lives are almost an afterthought when compared to the power of the devices themselves.

Don Ihde highlights that, in this pole of technological determinism, there are two important branches.[1] One, linked to utopias, suggests that technology by itself creates a better world, on that is more organized and free of pain and worries for human beings. On the other hand, there is the dystopian branch. New technologies bring with them the gift of the Greeks to the Trojans and, when fully welcomed and integrated into our cities, will reveal themselves as unavoidable sources of suffering and eventual destruction of societies and the environment.

At the other end of this continuum is a vision that we call technological neutrality. In this view, the impact of technologies is determined by its users, e.g., it is smartphones that are passive, virtually without agency, and how much they interfere in our world and our relationships depends exclusively on the choices users make, including how, when, and at what intensity to use them. From this point of view, digital technologies are ethically neutral; they are neither good nor bad in themselves. The ethical implications of these devices thus are entirely dependent on their usage.

Here we will pause to point out that in the digital age a technology's agency can vary widely. As we discussed in Part I, we willingly give agency to some apps. We don't want to risk missing a meeting, so we give an app the power to interrupt and remind us. Other apps have a kind of agency through their connection to other users. A text messaging app might not make decisions to interrupt us, but it will when triggered by messages sent by our friends and colleagues. This suggests that the landscape of this issue is far richer and more complex than is suggested by the two extreme positions of technological determinism and technological neutrality.[2] And so we now turn to an exploration of this landscape using important technologies currently in use, paying particular attention to theories that draw on elements from both poles. The theory of social constructionism is a key example, as it posits a two-way, circular relationship between technology and society. Building upon this more nuanced position, we examine how meanings are embedded into technologies that, in turn, change the way we live. Thus, there is a push and a pull in a cognitive-hermeneutic cycle, and is further grounded in the learning theory that we discussed in chapter six.

Like any cultural artifact, digital technologies build on an accumulation of meanings derived from culture, and in turn change how we see and interact with the world. Digital technologies are meaning-full, which is precisely why they are so relevant to understanding today's society. This is also why we believe that the cognitive-hermeneutic approach proposed here is necessary for investigating the deeper ways that new technologies impact everything from personal identities to interpersonal relationships to social institutions.

Assuming this cognitive-hermeneutic approach, the transformative power of technology is revealed by analyzing the relationship of society and digital technologies as it is particularly impacted by digital metaphors. Social networks are not merely applications used by millions of young people to pass the time; they redefine essential human experiences such as communication, truth, and friendship. Furthermore, the ways in which we experience the physical limitations of time and space are fundamentally changed in "cyberspace" through metaphors that re-signify human experiences. In this case, concepts such as "presence" and "simultaneity" change their meaning in a digital context.

Several studies have explored and emphasized the use of metaphors to create good user interfaces and experiences for digital artifacts.[3] Although these studies touch the same underlying phenomenon that we describe in this book, their intentionality is quite different from ours. They are largely inspired by pragmatic goals aimed at optimizing software engineering practices. In our case, we want to explore not only how digital metaphors can be used to bridge the conceptual gap that increases user engagement, but also the profound consequences they bring to personal, interpersonal, and social environments.

Our hypothesis is that the cognitive-hermeneutic approach

that we develop in this book provides new ways to analyze fundamental structural layers of the impact of digital technologies on society. Through their impact on their users' cognitive associations and semantic horizons, digital devices are shaped by digital metaphors that can radically transform how their users see the world, their personal relationships, and social, economic, and political phenomena. Our focus is not just user interaction with the digital artifact, but more importantly how a user's interaction with the world is mediated and reshaped by digital artifacts. These devices not only create new functionalities with a new set of possibilities to act in the world, they also affect the meanings that we attribute to the world, ourselves, and our relationships. Digital metaphors rewrite, at the semantic and cognitive level, things as fundamental as friendship, communication, social engagement, education, knowledge, and economic and political relations. They make us "see as," transforming human experiences at their most basic level.

Obviously, the plasticity of digital metaphors can lead to strong consequences, both positive and negative. For example, the fundamental metaphors of social networks related to friends' conversations have been reframed with new layers of meaning by malicious users intent on spreading misinformation and manipulating public opinion.

Therefore, the cognitive-hermeneutic approach that we propose touches the foundations of the complex relations between societies and digital technologies. Far from occupying a single point in the continuum, depending on context, digital metaphors' manifestations and applications come closer to technological determinism, but their extremely flexible forms of use and application can bring them closer to the arguments of technological neutrality.

The point we want to emphasize is that the cognitive-hermeneutic approach that we propose in this book allows us to analyze the more radical implications of digital technologies as they transform the world and individuals by changing the meanings in which we live. They are not just passive tools that we use to increase performance in certain areas of human experience; they redefine these experiences, sometimes with us as active participants. As Ricoeur proposes in his critical hermeneutic philosophy, all human action is mediated by meanings.[4] Basic human actions like chatting, cooperating, dating, marrying, doing business, and playing are always "symbolically mediated." They take place within a conceptual framework of the meanings we associate with each of them. As we have seen with metaphors, digital artifacts affect these symbolic mediations; they re-signify them and therefore affect the way we experience and act in the world.

Our approach allows us, therefore, to go beyond a functional analysis of digital technologies, enabling the exploration of their role in transforming the symbolic mediations of meaning that structure societies and individuals. By focusing on the hermeneutic and cognitive aspects of digital technologies, we touch a deep layer of the influence of such artifacts. Still, we do not want to exclude material conditions that are dialectically related to these aspects. Our discussions of the metaphorical spiral and technological enablers highlight the importance of an analysis that recognizes the complex interrelationships between the material-structural and cognitive-hermeneutic conditions of development and use of digital technologies.

As we will see in more detail in the next chapter, Coeckelbergh and Reijers have explored, using the concept of narrative technologies, the ways digital artifacts build structures of meaning that reorganize human actions.[5] Digital technologies in particular can reconfigure human time and mediate how people relate to everyday events. Coeckelbergh and Reijers cite the example of how electronic monetary technologies have redefined the mediations of meaning around trading actions.[6] The ways in which we understand and act on concepts such as hedge funds, options, securities, derivative trades, and currency swaps have been transformed by digital financial technologies. Human traders, in the wake of these digital technologies, have been thrown into a new universe of meaning in which symbolic mediations of actions are deeply affected and, in many ways, redefined by digital artifacts and architectures.

Also, we echo Johann Michel's proposal that we are fundamentally interpretive beings—*Homo interpretans.*[7] The symbolic mediations brought about by metaphors do not simply affect an isolated aspect of individuals and societies, they interfere in the essential mechanism of giving and interpreting meanings for the experiences in the world. It's not just about knowing differently, but living differently. Social networks not only affect what we mean by friendship, they further reframe this phenomenon and therefore restructure the ways in which we create, maintain, expand, and even break the interpersonal relationships of friendship. It is therefore not just an epistemological phenomenon—what we know about the world—but also an ontological phenomenon— what we are in the world.

In this section, we examine a number of the most important technology metaphors of the last two decades.[8] These metaphors vary in their effectiveness, both in terms of how well they make the technology understandable for users and how well they match what the technology is actually used for. As we shall see, metaphors' effectiveness will impact how fast they are adopted, and also how essential the hermeneutic cycle is. All of them, however, have significantly changed our lives, both what we do and how we think.

SPEAKING OF MEDIATING THE WORLD

The amount of agency digital artifacts possess can vary widely. We often willingly choose to provide them agency.

We use digital artifacts to mediate an increasing number of fundamental human experiences, such as communication and relationships. These mediations fundamentally alter key aspects of those experiences, thus changing their meaning.

CHAPTER EIGHT

METAPHORICAL FOUNDATIONS

Touching Is Selection

Our first example is also one of the most highly effective digital metaphors and concerns what may be the single most important software innovation of the twenty-first century, one so important and so ubiquitous that it is easy to forget that it exists and hard to imagine life without it: the touch user interface (UI). This technology first came to prominence when Apple released the original iPhone in 2007. In October 2020, only thirteen years later, Apple analyst Horace Dediu noted that "touch UIs are now enabling $1 Trillion of economic activity."[1] Touch UI is probably less important than the other metaphors we discuss in terms of meaning, but it is foundational to the rise of mobile technology as it provides a metaphorical base upon which all other mobile metaphors build. Just as "argument is war" is what Lakoff called a "structural metaphor" that serves as a framework for a variety of other metaphors, the basic actions of the touch UI system can be picked up by other digital metaphors and used without further explanation.

Before 2007, the cell phone industry was dominated by companies such as Nokia, Motorola, and RIM that either used hardware keyboards or a combination of a number pad and several dedicated buttons. This is the world Apple faced as it developed the iPhone.

For example, in 2005 the Nokia N70 became the first "smartphone" to sell more than a million units, and its UI consisted mainly of a number pad and a few other dedicated buttons. Meanwhile, the Motorola Q became the first "QWERTY" phone, one with an entire keyboard, to sell a million units in 2005. This market grew so quickly that by 2006 the Blackberry Pearl sold fifteen million units. At the time, the ubiquitous UI for cell phones was a combination of a list (e.g., a list of phone numbers in an address book or a list of applications to choose from) and the arrow keys on the hardware keyboard. Users used the arrows to navigate to the item in the list that they wanted and then selected their choice by pressing the "enter" key. The activity that made early smartphones "smart" was email, and the push to full keyboards came about because of the difficulty of typing emails on a number pad, a slow and tedious process. Cell phone keyboards themselves were essentially miniaturized versions of the keyboards used on personal computers. This had the advantage that many people already had experience with keyboards from computers, or even typewriters, and thus had a mental model of how they worked. As Steve Jobs recounted in his keynote introducing the iPhone, Apple saw this model as a problem that needed to be solved.[2] Aside from the mechanical problems possible in such hardware, the need for people to know how to type, and the difficulty in typing on such small keyboards, the problem Jobs focused on was that modern software needed to be dynamic in a way that existing interfaces could not handle. Jobs, after all, was involved in a many-years-long quest to remove peripherals from computers, a quest that would culminate in the MacBook Air's design, which included only a single USB-C port. The iPhone offered a real opportunity to ship a device without drives for external memory, one that could be updated without the need for physical disks. But updating software on a cell phone was hamstrung by the existing mapping to available buttons and the keyboard. If new software functionality was added, for example, there might not be any hardware buttons available that weren't already mapped to other things.

Once Apple decided to get rid of the hardware keyboard, its developers had many problems to solve, but certainly among the most important was simply how to let users choose what to do next. The operating system that Apple developed as part of this process, iOS, is built as a series of layers. At the bottom is what we would call the core operating system, which Apple refers to as Core OS. Each subsequent layer builds increasing functionality for those who want to develop software for iOS. At the top of this hierarchy is where user interactions are handled. In early versions of iOS, this layer was called Cocoa Touch. Apple's engineers had to create this layer as a bridge for app developers who want their applications to be interactive. Since the Apple developers writing Cocoa Touch could not anticipate every possible type of interaction that software developers might want to support, they had to create a system that was as general as possible. The simplest version of such an interface is simply to create a button for every possible interaction. For example, with a fixed number of actions, an operating system could just assign a button to each one. Alternatively, theoretically, with a keyboard a user could type in any necessary command, but this was a poor solution because the keyboards were so small. Instead, commands were normally presented as a list of items that users navigated with directional arrows. An improvement over this system, and the one employed by cell phones until then, was to create a list display that the user navigated with arrow keys. Once the user had navigated to the appropriate action, they could then take an action by hitting the appropriate other key. The metaphor employed by this system, such as it was, was "actions are taken by typing." Such a system is easy to implement as it can be repurposed for applications or actions within an application. On the other hand, it is not friendly to novices, requiring skills like typing that must be learned, and users could only see a few of the choices at a time, especially since the keyboards left little room for the displays. Removing the keyboard required a new way of thinking about this layer.

The obvious thing for the engineers tasked to solve this problem was to look at other models. At the time, Apple had reinvented itself largely on the back of iPods, a successful product that did not rely on keyboards but instead used a more intuitive interface based on a wheel. Users could choose one song among hundreds by spinning the wheel. The scroll wheel could also be used to choose applications or make selections, as it did on the iPod. (Note: Apple actually made mockups of iPhones using this design.) But the wheel had the major drawback: much like hardware keyboards, it interfered with the screen. If the screen were to be a central feature of the phone, then that left the engineers with little choice but to use the screen itself for the user interface. As it happens, around that time an enabling technology became available: the capacitive touch screen. With such a screen, it was possible to detect a user's fingers. Thus, the question became how to use this information to allow users to make choices. An obvious option, and one Apple explored, was to make a virtual scroll wheel—Apple even patented this technology.[3] In other words, in several different ways Apple could have easily relied on the knowledge that people already had of either cell phones or its own iPod. Instead, Apple went user-centric and asked how people choose things in the world. The answer is that they use their hands for grabbing and their fingers for pointing. As Jobs said in the keynote introducing the iPhone:

> We're going to use the best pointing device in the world. We're going to use a pointing device that we're all born with — we're born with ten of them. We're gonna use our fingers.[4]

Putting the technology together with how users do things in the world meant that a user could "point" at something on the screen, and the capacitive screen could identify where on the screen they touched it. Of course, the reason this layer is called Cocoa Touch is because the act of touching the screen became a driver for the entire interface. Thus, the metaphor "touching is selecting" is a bridge from the engineers' technical questions surrounding connecting software actions to user decisions and the user experience of learning to do things without needing new technical knowledge.

Previously, hardware buttons were dedicated to specific actions. With touch, it was possible to create virtual buttons for essentially any action, which in turn meant that software could change dynamically in the way that Jobs envisioned. As we have already seen, frequent software updates are a hallmark of digital technologies in today's world. This necessitated a new development paradigm in Cocoa Touch to let app developers put their own buttons on the screen and connect those buttons to arbitrary actions. In software terms, this paradigm is called event-driven programming. Cocoa Touch monitors the screen and can process what the user does, where on the screen the user touches, for how long, etc. App developers can grab the results of this processing and use it to determine the user's intent. Thus, Jobs's idea that software should be dynamic and upgradeable was now realizable because the general-purpose touch interface could support any sort of user interaction that app developers might dream up. Apple's choice to look to the user for inspiration instead of engineers led to a new software paradigm and, not coincidentally, the most successful consumer product in history—and in doing so, they used a powerful metaphor. Freeing the software from the tyranny of dedicated buttons had other advantages. Developers were free to make their software's user interfaces look however they wanted. On a cell phone, updates could be done without the expense and time required to make physical disks; they could be done more frequently and at effectively zero marginal cost.

The vehicle of this metaphor is "selecting" or "choosing," and the tenor is "touching." This is a powerful technology metaphor because, from the point of view of the people that need to use it, it almost isn't a metaphor at all. As Jobs noted in his keynote, fingers are the greatest selection devices ever created. The richness of the metaphor does not come from, nor does it require, a major cognitive reorganization on the part of the recipient; it comes from the ways that the technology was able to employ the metaphor. All of the layers in iOS, and now Android, work to hide all of the details necessary to make this happen. The metaphor makes us see touching as selecting, and very effective metaphors such as this one reshape how we see the world. By doing so, they become transparent.

Another characteristic of highly effective metaphors is that they become what Lakoff and Johnson call a structural metaphor, like "argument is war."[5] Such metaphors not only make us see arguments as war, they create a metaphorical region in which other connected expressions flourish, such as "Your claims are indefensible," and "His criticisms were right on target." In terms of touch UI, "touching is selecting" led to a series of related metaphors, including "grabbing is touching and holding," "swiping is moving," and others, creating a constellation of related metaphors and providing a foundation for other tech metaphors to build upon.

The touch UI metaphor has another layer that also contributed to its success. Touching an object to select it, or touching an object to "grab" it and then dragging it with one's finger—these are operations that have physical analogs. We touch things to select them, but we can also use our fingers to manipulate objects. Touch UI enabled software to behave in the same way. When Jobs first showed "slide to unlock" and "swipe to scroll," the audience audibly gasped. They may not have understood anything about how this technology worked in terms of capacitive screens and the software involved, but it was self-evidently useful and a leap ahead of what they were used to.

In retrospect it is easy to feel that the iPhone's success was a foregone conclusion, since touch UI is so obviously superior to hardware keyboards and other previous alternatives. And while it is true that the iPhone was hotly anticipated, that anticipation was built mainly on Apple's success with the iPod. Indeed, the touch interface was seen as the biggest obstacle to the iPhone's success. In one famous interview, then CEO of Microsoft Steve Ballmer laughed at the outrageously expensive device, noting that without a keyboard it was "not a good email machine" and thus would be shunned by business customers. Note how Ballmer's view of a smartphone's success was predicated on its use as an email machine. To use a hockey metaphor that Jobs was fond of, Ballmer was skating to where the puck was, in this case email, while the iPhone was skating to where the puck was going. In fact, predictions of the iPhone's failure were rampant upon its release, with news organization after news organization declaring it dead on arrival.[6] Even in 2009 articles were still being written about the iPhones "Achilles's heel" and how a keyboard should be added. In many ways the tech press's reaction to the iPhone reveals the disconnect between people within the tech industry and the general public. People within the industry are "power users" and thus are not representative of the public since their mental models of technology are so much more developed than those of the average person. Such users are interested in tricks and shortcuts, wanting to maximize efficiency. The public, on the other hand, just wants something that is fun and easy to use without requiring a slow learning process. The iPhone became the most successful consumer product in history by meeting these needs, and much more. It is worth noting that this pattern has repeated itself over and over with Apple products; iPads, Apple Watches, and AirPods were all derided by the tech press upon release, and all went on to dominate their markets, in no small part because they were built on metaphors that made them easy to use and understand.

The iPhone sold six million units in 2007, and the impact of touch UI on the rest of the industry was immediate. By 2008, when it sold twenty-five million units, there were already two other phones that used touchscreens and sold more than ten million units, one from Nokia and one from Samsung. By 2009, phones running Android, Google's operating system that was a competitor to iOS but also used a touch UI system, started to appear on the market. Nokia was able to keep releasing phones featuring hardware buttons that sold in the hundreds of millions until 2015, but by that time the smartphone market had essentially been completely taken over by iOS and Android. By the first quarter of 2011, Apple earned more profit in the phone market than the rest of the industry combined, while Android was described as a "lifeboat" for the rest.[7] It took just over three years for touch UI to completely overwhelm the smartphone market. This transition was able to happen so quickly in part because of the cognitive-hermeneutic cycle. The iPhone was, and still is, a high-end product. As late as 2013 Nokia was writing about the importance of making cheap smartphones on its company blog.[8] But companies like Samsung were able to take over that part of the market because Android allowed them to make better phones cheaply as well. Android was essentially free, and its software made phones that used it capable of an almost infinite number of things.

In the meantime, "touching is selecting" is a metaphor that has become familiar to everyone who has used a smartphone or tablet in the last decade, including a new generation of children who never learned any other way of interacting with them. Stories of such children trying to touch and swipe other devices like televisions are legion as users increasingly have become unaware of the layers of technology required to make it work.

It is worth noting that the metaphor is essentially unchanged more than a decade after its introduction. There have been incremental improvements to how the metaphor is implemented, but not the kind of full-fledged hermeneutic cycle that we discuss throughout this book. For example, there are new touch gestures added periodically to both the Android and iOS systems, but most of these are mainly useful for power users. The vast majority of user interactions with phones come from touching, dragging, tapping, and swiping, all gestures introduced in the first iPhone. There have been experiments to modify the experience, such as Apple's introduction of force touch, where the level of force that the user applied could differentiate the operation selected. But such complications do not improve the metaphor; they muddle it, and none of them have found mainstream use. It is reasonable to assume that many of these variations have not found widespread use because they simply do not match the physicality of the original metaphor. Users of the first iPhone would find it simple to adapt to the latest phones by any manufacturer. This is mainly due to the strength of the original metaphor and the level of match to the technology involved.

The real importance of touch UI, though, is that it is an enabling technology that provides a foundational layer that other mobile applications can build upon. In 2021 the publication *Inc* called Steve Jobs's email giving developers approval to write apps using Apple's touch UI software "the most important email in the history of business."[9] Consider web browsers. The lifeblood of the internet comes from the hyperlinks embedded in web pages. A user on an e-commerce site such as Amazon can easily scroll through a page using their fingers and then touch and select any item they wish to buy, or click on any links to other pages by tapping them. It is nearly impossible to imagine how this would work on a phone without touch UI. Meanwhile, since every app uses touch UI in their own interfaces, they are free to build on it, exploiting the fact that users will know how it works. Apps do not normally need to teach users about selecting and swiping because that is what they are used to, but the generality of touch UI is also such that a gesture like a swipe can be easily repurposed by individual apps, as Tinder famously did when it built its selection mechanism around swiping either right or left to show one's preferences.

SPEAKING OF TOUCHING IS SELECTING

Touch UI is perhaps the quintessential example of the value and power of metaphor in technology. "Touching is selection" can be effortlessly understood and thus is transparent even to a new user. Meanwhile, the metaphor hides a complex set of technologies that users need not be aware of. As the primary mediator between phones and users, touch UI also serves as a metaphorical foundation for apps to build upon.

CHAPTER NINE

MEDIATING RELATIONSHIPS

Social Media

Other important innovations that came with the introduction of the iPhone was the app store and the subsequent "app economy." In the early stages of the personal computer revolution, software was expensive and difficult to update. It came on physical disks and required a significant investment of time and money, both on the part of the producers of the update and on the part of the consumer. By the time the iPhone was introduced in 2007, this was starting to change as computer applications could be downloaded over the internet. This was a huge advantage for developers, because it reduced the costs of updates to virtually nothing, and for consumers, because they could get software more cheaply, did not have to go to a store or wait for mail, and update the software at any time over the internet. But even with the ability to remotely update software (over-the-air updates), only device manufacturers could use this functionality. So, all mobile device software was created by a limited group of developers associated with hardware manufacturers. This ecosystem limited both the creation of new applications and diversification of experiences. As we shall see in chapter twelve, even now companies like Apple who have profited greatly from app stores are still reluctant to completely open up their ecosystem.

The app store represented the next step in this revolution. From the point of view of developers, a new app they created could automatically go to the same "store" as every other app, and every customer in the world could potentially obtain it. From the consumer point of view, apps could be added on a whim. Over time, the fact that there are no marginal costs to apps after their development led to a collapse in the cost of software. Whereas previously a typical computer application would run at least fifty dollars, and often into the hundreds of dollars, the typical price of an app in the mobile market quickly became "free." This led developers to look for other ways to make money and naturally led them to look at advertising.

With the proliferation of apps in the era of mobile phones and the subsequent drop in the price consumers are willing to pay for them, most of the money made by developers in the modern app economy comes from advertising.[1] To succeed, then, an app's software development must gain and hold users' attention. For software developers participating in this "attention economy,"[2] this problem can be framed as "how can we keep people engaged (online) as much as possible?" From this point of view, software that competes in this attention economy is attempting to supplant previous industries that also competed for attention by providing entertainment, such as movies and television. Indeed, television itself is built on the metaphor "a window to the world."[3] Apps as a whole are effectively working to replace television as that window. The previous solution to holding the public's attention, as embodied by the television industry, was for the industry itself to produce as much content as possible. Of course, creating such content is expensive and risky—after spending large amounts of money developing a show, television networks would essentially hope that the show was what the consumers wanted, and many such shows were dropped after only airing a few episodes. Thus, the temporal dimension of the old model was long; creating a new show, writing it, producing it, and finally airing it could take a year or more. Nevertheless, the power of this model is such that not only does the television industry still use it, but so too do Netflix and more recent digital competitors like Amazon Prime, Disney+, and HBO Max, because control over premium content is one way to hold users' attention.

The rise of social networks represents an alternative model in which content actually comes from the users (we note here that the gaming industry affords yet a different model). This solves two major problems simultaneously: app developers are no longer responsible for the creation of all of the content required to keep users engaged, and content is likely to be interesting and engaging to users because they are creating it themselves. Such content can also be far more responsive to what is going on in the world, since that content can be generated very quickly. In other words, where television was a metaphor for a window to look out at the world, social media is a metaphor where the world can look in your window. For this to work, these networks need as many people on and connected to each other as possible to keep the supply of content high. Thus, it is in the best interest of developers who work on social networks to design their software to encourage people to connect with each other even if they have no relationship in the real world. The collective connections between users are sometimes called the social graph, and the idea from the developer's perspective is for that graph to be densely connected. Thus, in this reconfiguration, users of social networks are encouraged to form relationships with people they may never have met. Facebook "friends" may never even have been in the same room. "Conversations" with celebrities and people on the other side of the world are equally possible. Further, the temporal dimension of content creation has shrunk to almost nothing—content is consumed almost immediately after it is created, and the consequences of poor content are minimal.

Social networks go beyond the simple "screen as entertainment" metaphor, though. The social graph is built around the semantic context of relationships and thus refines the general "entertainment" metaphor to be more specifically about relationships, be they friendships, love, or other. (Because they want to keep you engaged, networks like Facebook have grown to incorporate other features as well.) These relationships are expressed somewhat differently in different networks, but in general social networks attempt to capture the spirit of checking in on or catching up with existing friends, or meeting potential new friends or starting new relationships. In the predigital world, friends interacted by meeting in person, which required being in the same place at the same time. When people met with friends, relatives, and acquaintances, they inevitably spent time trading information, or "catching up," telling each other what had happened in their respective lives since the last time that they had seen each other. In a social network, a "post" is a metaphor for this process. A post, be it a tweet, a picture, or a video, informs the world what the poster has been doing or thinking. In a real-world, or predigital, situation, our friends respond to our news, creating a conversation and providing support. In a social network, the metaphor turns these conversations into comments on the post, and support is expressed and collected in the form of "likes." Thus, the full metaphor: "catching up with friends is posting, commenting, and liking."

A given social network contains "feeds," which again build on the foundation provided by the basic software list construct. Every time you check in with a social network, it presents you with the posts that have been made since the last time you were on. The simplest organizational structure for such feeds, and one most easily implemented with a list structure, is a list ordered by time—a timeline. And indeed, many social networks, such as Twitter, are organized around this concept. However, a raw timeline would break the catching up metaphor and could be confusing as the number of posts grows. For example, in 2018 technology writer

Benedict Evans estimated that the average Facebook user would have between 1,500 and 2,000 items in their feed each day based on 150 friends posting or linking 10 to 20 items per day.[4] This shows just how well Facebook has solved the content generation problem, but also why the feed needs additional organization. So user interfaces in social media are generally conceived as a combination of time and conversations. The result is a system that is more coherent and meaningful than if users just saw a long list of items ordered by time. This additional structure is a necessity, because the technology removed the previous restrictions on conversations that are fixed in time and space. The extra structure also provides the user with a mental model that allows them to make sense of what otherwise might be a string of seemingly unrelated conversational snippets and to overcome the semantic distance from the concept of conversations face-to-face. The conversation metaphor also allows users the illusion of seeing people they are connected with as friends. Thus, the refiguration of the relationship concept under this metaphor not only breaks temporal and spatial boundaries, it even stretches the definition of relationship to include people that we may never have met.

The "friends catching up" metaphor is a useful starting place for a social media company. It is familiar to everyone, it takes advantage of the connected structure of the internet, and—with the concept of a feed organized by a combination of time, people, and conversations—it is easily grasped by even nontechnical users, who are not required to learn specialized commands or to use complicated interfaces. Meanwhile, by expanding the metaphor beyond one's close friends and relatives to include even people that the user may never have met, social networks have expanded the semantic field of what it means to catch up. On one hand, people can connect to an arbitrary number of other people whether they directly know them or not. On the other hand, the information that they choose to share with others can be carefully curated if desired. The intimacy of close sharing with a few friends shifts to public-facing sharing with as many other people as wanted. In addition, catching up with friends usually takes place in specific contexts, locations, and predetermined times that make what is shared restricted to the context of who is being spoken to and when. By reframing "catching-up," social media timelines break these temporal and physical boundaries, creating what dana boyd called collapsed contexts, which profoundly alter the dynamics of self-expression.[5] This weakening of what it means to be a friend directly benefits social networks, as it helps them solve their content creation problem in the attention economy.

Meanwhile, the cognitive-hermeneutic cycle continues as social networks compete with each other. As of this writing, the currently hot iteration is TikTok, and even newer audio-based networks such as Clubhouse. At a glance it might seem like TikTok is just another social network where users create the content. However, this view ignores a key factor in TikTok's success. The constraint of a pure social network is that, while users create the content, you are still relatively constrained by only seeing the content of people that you follow—your friends. Social networks continue to tweak this idea, e.g., Twitter allows users to retweet others, a version of "a friend of mine said . . .", thus leveraging the power of the network. TikTok removes this constraint altogether in favor of its sharing algorithm. Now users essentially have access to the content created by all of the other users. This is possible in the abstract on social networks for items that have enough viral reach, but TikTok removes the friction in the process that comes from the network. In general, TikTok's success can be traced to removing friction that way, and also in reducing the friction of making new videos, which in turn helps ensure a large supply of video content.

Feeds, whether TikTok's algorithmic feed of videos or Facebook's newsfeeds, have evolved in a kind of arms race where the goal is user engagement. Ultimately, because of this the competition, apps require high volumes of content that they try to tailor to individual desires. As tech writer Benedict Evans put it, "all social apps grow until you need a newsfeed. All newsfeeds grow until you need an algorithmic feed."[6] TikTok and Facebook make no bones about the fact that their algorithms are designed to optimize engagement. Engagement here is a proxy for attention. Recall from chapter six that we as individuals experience a kind of constant tug of war between our inclinations and our intentions through voluntary and involuntary attention. Voluntary attention is associated with will and is deployed to help us reach our goals. The feeds of social networks are designed to be so distracting as to thwart our will to do something different or to be productive. There is always another cute cat video or another political story designed to provoke outrage. Since there is an ecosystem of such apps, their inevitable solution is to deploy machine learning to determine exactly what your needs are and then draw on their vast store of posts to provide that to you. Framed another way, these apps are learning how to thwart your willpower, and every time you hit a "like" button or a heart icon you are teaching them how to do it. As Adam Alter's book title suggests, timelines become "irresistible" as they tap into behavioral addiction, and effective digital metaphors can be one crucial component of such behavioral engineering. Alter mentions Tristan Harris's comment that it is not that people lack willpower, but that "there are a thousand people on the other side of the screen whose job is to break down the self-regulation you have."[7]

A great deal of the writing on the topic of the information about users that these companies are acquiring is usually framed in terms of advertising, and it is indeed important since it provides revenue, but there is no one to advertise to if they have abandoned your app for your competitor. While it is true that users are not the customers of these apps, the apps need as many engaged users as possible in order to survive.

Before the virtually infinite choices of the internet and machine learning algorithms deployed to hold our attention, in 1985 Neil Postman wrote a prophetic book, *Amusing Ourselves to Death*, in which he described how the medium of television was designed to hold our attention and the rather dire (from his point of view anyway) consequences of a nation seeking and finding constant amusement as television undermined all other forms of communication. In the foreword, discussing the differences between *1984* and *Brave New World* he noted:

> Orwell feared those who would deprive us of information. Huxley feared those who would give us so much that we would be reduced to passivity and egoism. Orwell feared that the truth would be concealed from us. Huxley feared the truth would be drowned in a sea of irrelevance.[8]

Postman noted that his book was about the possibility that Huxley's *Brave New World* was right. There is little reason to believe that the subsequent decades since the book was published would change Postman's mind. When Postman was writing, most people had only four channels of television to choose from. The digital world, on the other hand, affords almost infinite choice, including many products that have relentlessly refined the art of capturing and holding attention. As we noted in chapter three, Herb Simon, writing at about the same time as Postman, called attention a "precious resource" in an information-rich world.[9] Fifty years later the phrase "information rich" is a dramatic understatement a digital world where there is more information than ever, but also where human connections are being increasingly diluted.

SPEAKING OF SOCIAL MEDIA

Social media apps seek to mediate how we relate to each other through metaphors based on concepts like friendship, dating, and social groups.

The aims of the developers are to win and hold the attention of users. Their methodology mainly relies on the constant develop-

Table 1: Different aspects of relationships with and without social media

Relationship Features	Raw	As Mediated by Social Media
Meeting	May be serendipitous, may be through mutual friends	May be discovered, may be sought
Catching up	Talking in person	Posting, reading posts
Support	Providing counsel, encouragement, advice, etc.	Liking posts, commenting on posts
Intimacy	Varies according to level of friendship, but leans towards more personal interactions	Can vary, but there is a strong pull towards being public-facing

ment of content by the users themselves. Positive framing metaphors, e.g., "like" buttons, aid in this task.

Users are able to maintain relationships without the distance and time constraints imposed by in-person interaction.

These apps are learning how to thwart your willpower and every time you hit a "like" button or a heart icon you are teaching them how to do it.

CHAPTER TEN

MEDIATING COMMUNICATION
Texting

A common property of mobile applications is that they provide the ability to break spatial and temporal boundaries for activities that were once constrained in either or both dimensions. The fact that smartphones are still referred to as "phones" gives primacy to their role as phones that are no longer tethered to a single location. Given this fact, it is natural for the mobile phone to have begun an era of evolution in how people communicate. A number of communication alternatives to smartphone-as-mobile-telephone have sprung up in the wake of the cell phone revolution, perhaps the most important of which is texting. The calling apps on cell phones removed space as a constraint on communication, but not time. Making a call no longer required you to be in a specific place, tethered to a wire. Nevertheless, to talk to someone with a mobile phone, you are still required to use the calling app at the exact same time that they do—you must be temporally in sync. This suggests that apps that remove the temporal constraint would be in high demand. The antecedents of apps which do this, text messaging apps, are simple—telegrams and letters. Both removed spatial and temporal restrictions on communicating; one need not be close in time nor in space to communicate with someone else by telegram or by letter. That telegrams, and especially letters, were such a strong influence on the development of texting can even be seen in the iconography of many text applications, which resemble envelopes for letters.

Texting dates back to 1984 and a system called "SMS," which stands for "short message service" and initially limited texts to 160 characters so they could fit into existing signaling formats.[1] In other words, the original version of SMS was shaped almost entirely by a combination of the preexisting forms of communication that removed temporal and spatial restrictions, along with the technological limitations present at the time. Thus, SMS did not start out as a reimagination of communication; rather it began as a very straightforward marriage of technology to an existing metaphor. The SMS format would later lead Twitter to adopt the same character limit for tweets. The original metaphor "SMS messages are telegrams" is a kind of second-order metaphor in that a telegram is a metaphor itself for communication, thus it does not resonate as easily as a metaphor like "touching is selecting."[2] By contrast, "tweet" stands on its own and captures the brief nature of messages on Twitter. Note that we use telegrams in the original metaphor mainly because of the length limitations. Telegrams were written to be short; the advantage of letters is that they can be as long as the writer desires. The choice to use letters in the iconography of early SMS apps is undoubtedly related to the fact that virtually everyone has experience with mail, but the use of telegrams is relatively rare in the modern era.

The strong ties between the original formulation of SMS and technology hampered its adoption. To a user without expert knowledge, the 160-character limit is baffling, mysterious, and completely unnatural with respect to familiar forms of communication. Then there were the problems of actually creating a message. As we saw in chapter eight, early phones did not feature full keyboards so typing in a message involved a laborious process that was not intuitive to nontechnical users. In other words, the technology itself was a barrier for nontechnical users, preventing them from fully acquiring the metaphor.

Over time, many of the original technological limits on SMS and its usage were dropped one by one in an ongoing technical-hermeneutic cycle—keyboards were introduced, autocomplete technology was added, texts could be longer than 160 characters, messages could eventually include audio and video, etc., in a kind of constant reformulation of the technology. As this happened, developers worked to incorporate the changes into their systems in order to improve the software and make it more useful. These changes naturally broke the original, clumsy, metaphor, which eventually transitioned to "instant messaging." While letters remove one kind of temporal restriction on communication, they are anything but instant. By emphasizing the speed of SMS, developers were able to reformulate the original metaphor.

The new metaphor became "texting is conversation." This is an improvement over the original metaphor in many regards, not the least of which is that it is no longer a second-order metaphor. Unlike in social networks, where posts are made to everyone who is a follower, in a text message thread, conversations are had with selected people and thus are more focused. And, unlike a letter, a text message thread between two people online at the same time can be in real-time the way an actual conversation is. There are even a number of ways that texting does not simply recreate talking through technology, but actually improves the original experience. We have already noted that text conversations do not have the same temporal and spatial constraints that real conversations have. In addition, text conversations have reduced attentional demands that can allow for multi-tasking (or even having multiple conversations simultaneously). If desired, these conversations can even leave a concrete trace. They do not have to be remembered, they can be scrolled back to; conversely they can also be deleted, and some apps will do this automatically so there is no "paper trail." Finally, they do not require speaking, meaning they can happen even in places where speaking aloud would bother other people nearby, although the advent of audio texting meant that spoken texting is possible. Thus, texting has expanded the meaning of what a conversation is.

By 2010 SMS had become the most widely used application in the world, ubiquitous to the point where calling mobile devices "phones" does not really make sense anymore, something that we explore in chapter ten.[3] (The usage of SMS specifically is no longer as high, as it has increasingly been replaced by similar systems such as WhatsApp and Apple's iMessage that add their own functionality.) We showed in chapter two how deeply the "texting is conversation" metaphor has become embedded in young people. Further, texting, especially as it has evolved beyond SMS, continues to further untether the rules of conversation. Originally texts were sent to a phone number, which in turn was connected to a specific device, just as phone calls had long worked. Now with iMessage, for example, texts can be sent to a person and can appear on any of their registered devices. In turn, this has even allowed companies to modify calling apps so that they work the same way; thus, you can receive phone calls on your laptop or your tablet. The metaphor has outlived and evolved beyond the original technology for which it was developed. This is an example of how a digital metaphor can be more important than the specific technology it is initially paired with.

It is interesting to consider the impact of texting in terms of demands on our attention. It might seem that texting is actually less demanding of attention than a face-to-face conversation. Looking closer, however, it is far from clear that this is the case. For one, for many people new texts are typically accompanied by notifications, which we have already noted have a significant cost in terms of attention. Responding to texts is also likely to break the flow of one's activities into smaller and smaller chunks, requiring a constant shifting of attention from one thing to the another. Each such shift is likely to be accompanied by a need to reorient and refocus on one's previous activity. This wouldn't be the case with most face-to-face conversations. People and talking are both things that spur involuntary attention. It does not normally take much attentional effort to converse with other people because we are wired to find other people interesting. Typing, on the other hand, does require focused attention. There are variations, of course, depending on circumstances, but casual in-person conversations should not significantly drain attention.

Many people have reported a further problem of attention and texting—that they experience anxiety when waiting for texts. In *Alone Together*, Sherry Turkle reports on her research with many teenagers who wait anxiously to receive responses to their text messages, particularly if they involve emotional content.[4] This anxiety is distracting and stressful and may even be alienating to nearby people who notice the constant furtive glances at one's phone. In a face-to-face conversation you rarely have to wait for the person you are conversing with to reply. There are occasional awkward silences, of course, but that very awkwardness generally spurs people to talk. In a texting conversation, in which participants are freed from looking at an expectant face, such silences can stretch for long periods of time. And the silence can mean virtually anything because, as we just noted, people can easily multi-task while texting and thus might simply be engrossed in something else. Texting has many potential benefits for the spatial dimension of conversations, but the loss of seeing the other person and the huge amount of context that comes along with facial expressions and gestures should not be understated.

SPEAKING OF TEXTING

Texting began as a technologically limited metaphor based on a hybrid of telegrams and mail. As the technology improved and

Table 2: Various features of having a conversation in person versus through texting

Conversation Features	Raw	Mediated by Texting
Ability to discern intent	Excellent, can use inflection, facial expressions, gestures	Depends on the skill of the person writing the text. Can use standard abbreviations, emojis, gifs, etc to augment
Scope	Normally lasts as long as the participants are together	Can take place over long periods of time
Permanence	As with any memory, degrades with time	Can be effectively permanent or ephemeral based on user choice
Conventions	Learned over a lifetime of practice	May vary widely.
Timing	Participants take turns. Pauses normally indicate it is time for someone else to speak.	Sometimes a text signals it is the other person's turn. Sometimes it doesn't. Pauses may be indicative of virtually anything.
Privacy	Depends on location	Depends on the app.

usage grew, the technology evolved and left the original metaphors behind. The new metaphor is that texting is conversation.

As with social media, texting drops previous time and space limitations on conversations. Part of the evolution of texting involves trying to overcome the limits of text. Examples include adding the ability to include media and the continuing development of emojis.

CHAPTER ELEVEN

MEDIATING MEMORY

Photos and Videos

Nearly every aspect of photography has been disrupted by the mobile phone industry. The old aphorism "the best camera is the one you have with you," which is attributed to Chase Jarvis, points to why. Having a camera with you no longer takes forethought, nor extra space, nor film or other equipment—it comes integrated into a device that most people already have in their pockets. In other words, smartphone cameras are convenient, and the smartphone industry has clearly shown that convenience is a powerful force in consumer preference. In addition, the cameras on mobile phones are increasingly able to challenge all but very expensive standalone models for quality, especially with the rise of computational photography enhancing their capabilities. Camera quality has become a differentiator for people buying new phones. The concept of having a phone with a camera is an old idea founded on the limitations of landline phones. As we discussed in the last chapter, while phones removed spatial restrictions on our ability to have conversations, they came with the cost that you could not see the person you were talking to. Thus, it was easy to imagine adding a video component to phone calls that would mitigate this loss. The idea long seemed like such a natural evolution of the phone that Dick Tracy even started making video calls with his watch in comics strips starting in 1946 (and notably, this technology is now common too).

As more and more people got mobile phones, the cameras on those phones got better and better, and storage improved commensurately, it became more and more possible for people to record their lives. This coincided with the rise of the "selfie," a picture taken of one's self and the *Oxford English Dictionary* 2013 "word of the year." The evolution of the selfie is closely tied to social media—the word began as a tag to mark what kind of photo it was. As we discussed in chapter nine, a function of social media is to catch up with friends virtually. Selfies are a simple and natural way to enhance this process; they are an easy way to show what one is doing and where. If a picture is indeed worth a thousand words, then a few selfies can be used to create an effective narrative without the time and effort of typing.

In our discussion of meaning in Chapter 1, we noted that even the most basic functions of our lives, such as eating and mating, are mediated and symbolically interpreted. This is true of our memories as well. As we saw in chapter 2, learning is impacted by repetition, emotions, and other factors. Since memory is just a specific type of learning, it stands to reason that memory has the same characteristics. Further, as we saw with learning more generally, memories will be impacted by Hebb's rule and thus will change over time, e.g., as we recount stories of the past to our friends, those retellings will actually affect the memories. It turns out that when our friend's memories are different from ours, those differences impact our own memories too.[1] In Chapter 2, we framed learning as mainly serving to help us make predictions, but it is also important in that it provides a link to our past. Thus, many ceremonies surrounding death revolve around memories of those lost to, among other reasons, signal that we too will be remembered after we are gone.

In many ways, the history of technology is one of providing ever better ways to capture our stories and moments in our lives. Arguably one of the very first uses of technology, drawing on cave walls, was an attempt to capture and communicate a memory. The invention of writing represented another milestone in preserving our stories and memories. As technology has developed our ability to do this has commensurately improved, and this process of improvement has seen an exponential leap in the past decade or so.

People have been "recording" their lives through diaries, letters, and the like for centuries. Like so many other things, digital technology has only made this easier. A diary, for example, requires time and discipline. Taking a selfie with a device that is always with you, on the other hand, only requires scattered moments here and there. One of the first responses of the app economy to the rise of selfies was to create social media services, such as Instagram and Snapchat, that focus on pictures. Suddenly diaries required little or no writing at all. Simultaneously these services and others provided easy ways to edit and manipulate photos, a skill once consigned to darkrooms and specialized training and equipment. Thus, especially in the context of social media, the life that was being recorded could actually be enhanced. The face that people started showing to the world could be altered digitally. For people eager to get more likes on social media, selfies evolved from casual snapshots to carefully curated and edited photos designed more to project a desired image than to accurately reflect reality.

Thus, there was a kind of industry-wide technological-hermeneutic cycle in place with phones and users. The possibilities that phones afforded caused app developers to respond with new digital technologies. In turn, these new technologies began to change users' behavior, e.g., taking selfies drove the app economy in new directions. An important moment in this cycle occurred in 2013 when Snapchat introduced a new feature called "Stories" where users could string a series of photos together into a single post. Like "touching is selecting," in retrospect the idea of stories seems so natural that it is almost hard to imagine life without it. Also like "touching is selecting," "a series of pictures tells a story" is almost no metaphor at all. Since its inception the concept of stories has been borrowed or stolen by nearly every tech company that deals with photos. Instagram essentially copied it directly, debuting its version of Stories in 2016. Photo storage services, such as Apple and Google, which clearly were not going to keep the ephemeral aspect of stories (on Snapchat and Instagram a story disappears after twenty-four hours), transformed the idea into "Memories," borrowing from a service called Timehop. The idea of Timehop was to periodically remind us of what was happening in our lives in years past by showing us our social media posts. It was natural to combine this idea with photos. Google has cited the idea that in tracking user behavior on its photo service it noticed that many users never looked at many of their photos. Thus, its Memories features is designed to be ostensive: to put photos in front of users whether they asked for them or not. Thus, Memories, like Timehop before it and the notifications from calendar apps, is another example of the intrusiveness of technology no matter how benign the motivation.

The metaphorical use of memory is clear in the way these applications explain their basic functionality to users. For example, the Google Photos application's help page says "Memories are collections of some of your best photos and videos whether from previous years or recent weeks." It adds the note (which becomes almost ironic when we think metaphorically) that "Memories are available on Android devices, iPhones, and iPads."[2]

Both Apple and Google build Memories algorithmically, choosing a selection of photos and videos based on when and where they were shot, as well as selecting based on computationally determined measures of quality. The idea, as is often the case, is to make users more engaged with the product. For users the appeal is convenience. The ease of shooting photos and videos also increases the sheer volume of content that must be stored and organized.

What good, after all, is a great photo if you never see it or cannot easily find it when you might want to see it? This is yet another variation of the timeline problem that social networks faced, and algorithms are again the solution. Of course, just as they are with social networks, algorithms can be problematic. Capturing memories and reminding us of painful experiences was a very early side effect that continues to this day, and one that was amplified by the COVID-19 pandemic.[3]

There is another, more subtle issue with intrusive memories—they can, and do, change our remembered experiences. In an article in *The Atlantic*, psychologist Robyn Fivush notes that:

> We use our memory in part to create a continuous sense of self, a 'narrative identity' through all of life's ups and downs: *I am a person whose life has meaning and purpose. I'm more than the subject of brute forces. There's a Story of Me.*[4]

More than ever that story is subject to changing and editing, whether by the selection and alteration of digital photos and videos, or the fact that tech companies choose what memories to highlight and how to package them together. Meanwhile, these digital encounters do alter our memory. As we saw in chapter six, memory is closely linked to repetition, so the photos that we see repeatedly will naturally become more and more familiar to us, coming to the foreground of our memory of an experience, potentially at the expense of other details. Even the most intense memories, so-called "flashbulb memories" associated with traumatic events, are altered through recall.[5] This is yet another side effect of Hebb's rule and the fact that it is functioning whether we are experiencing something for the first time or reminiscing about it.

With Memories, both Google and Apple provide agency for the motivated user, but remove the need for agency for everyone else. Both services allow the user to edit the created Memories or to create new ones themselves, but ironically the algorithmically gen-

Table 3: Characteristics of memory with and without digital mediation

Memory Features	Raw	Digitally Mediated
Fidelity	Poor to good, vivid details are generally filled in rather than remembered	Excellent
Scope	Poor to great depending on the uniqueness and importance of the memory	Limited to what is captured
Permanence	Most degrade with time	Do not degrade with time, but may become inaccessible with loss of support depending on format
Malleability	Memories change every time we recall them, more so if other people are involved	Can be tuned and crafted as desired

erated Memories only reduces the motivation to do so for all but a few users. This is a fairly typical effect of technology; by making so many things so easy and so convenient, it makes anything less convenient less appealing. Once combined with social media, this can be a potent combination. A small number of highly motivated users, so-called "influencers," can show off their highly edited and curated lifestyles to their followers. In turn, those followers cannot help but compare what they are seeing on social media to their own lives.

SPEAKING OF MEMORY

Photographs and videos began as a simple add-on to smartphones. As part of the metaphorical spiral, and especially in conjunction with social media, the usage of cameras evolved to provide a kind of record of one's life.

Recently tech companies have been connecting photographs and videos to the concepts of stories and memories. These narratives can be easily created and are increasingly done so automatically by artificial intelligence. They can also be edited.

It is the nature of memory that any time we engage with our memories we simultaneously alter them. Thus, our digital memories are altering our real ones.

CHAPTER TWELVE

MEDIATING THE SELF

Smartphones

In the tech world there has long been an obsession with a metaphor that is slightly different than most of the ones discussed in this book. The metaphor involves finding the "killer app," an app that is so obviously useful and necessary that it makes a new gadget or service a must-get. The term first appeared in *PC Week* in 1988, which explained that "everybody has only one killer application. The secretary has a word processor. The manager has a spreadsheet." Indeed, the word processor and/or the spreadsheet are widely considered to be the first killer apps for early computers, especially since they accelerated the adoption of computers in business settings. Early home computers were mainly aimed at those who might today be called "makers," tinkerers who enjoy putting things together.

When Steve Jobs introduced the iPhone, he was keenly aware of the need for a killer app given his experience founding and running Apple Computer Company (now Apple Inc.). In the keynote speech introducing the iPhone, Jobs said "the killer app is making calls." His use of "killer app" was aimed at the press who regularly used the metaphor as a framing device at that time; the reminder that "this device is a phone" was a metaphor for the public. Jobs undoubtedly knew better than anyone that the ability of an iPhone to make calls was not what made the iPhone a great device. But he also understood that to get people to buy an iPhone he had to convince them that it was a must-have in terms that they already understood. This should be a familiar refrain to the readers of this book by now. And, as we have noted throughout the book, the initial choice of a metaphor can have serious long-term consequences. By choosing a framing centering the device's ability to make phone calls, Jobs built a narrative that still has consequences more than a decade later. Very few people today would argue that making phone calls is an iPhone's killer app; many would argue that it isn't even an important one.[1] This is mainly because so many alternatives are now available, both on the device and off. On one hand, apps for email, texting, and other communications have made phone calls less important. On the other hand, internet services enable users to make phone calls on virtually any device, such as tablets, PCs, and watches. Further, you no longer have to send a text to a phone number. Instead, you can generally send texts to individuals, and your ecosystem will route it to the appropriate device—phone numbers have become incidental to what phones do. In 2014 when Jeff Bezos introduced Amazon's new smartphone, the Amazon Fire, he spent an hour and a half describing it and never once mentioned that it had the ability to make phone calls.[2] Other countries have not held on to this convention of linking the devices to phone calls—in Britain they are called "mobiles" and in Germany "Handys," each name emphasizing a critical feature of the device. Meanwhile, by 2019 it was estimated that the average iPhone had more than eighty apps installed, nine of which on average were used every day.[3] The killer app of the modern smartphone is its generality. Yet we still call them smartphones, and the notion of these devices as phones still persists and still impacts how we think about them.

Perhaps it seems like this does not matter; perhaps a smartphone is whatever one chooses to make of it. There has been online debate on whether things such as tacos qualify as sandwiches for nearly a decade now, and something as complex as a smartphone seems even less likely to fit into neat boxes, so why should we even try? It turns out that, among other things, there are strong legal implications and, as we shall soon see, other consequences as well.

An example of why this matters arose in 2020, when Microsoft wanted to expand its online game system to include iOS. Microsoft wanted its Xbox customers to be able to start a game on an Xbox and later pick it up elsewhere on another device. In other words, Microsoft wanted to remove location as a restriction on playing Xbox games, and, as we have seen time and time again, removing location restrictions is one major advantage of mobile devices. Apple in turn blocked Microsoft's app from appearing in its app store, stating that it would not be able to review all of the games that might be played through the app. Microsoft and its customers reacted with anger. How one views this situation probably reflects one's model of what a smartphone actually is. If you view it as a small computer, then Apple's move does not make sense. For most of us, computers are devices that we control and can configure how we like, installing whatever software we please. The Xbox, by contrast, is a console. Users understand that its function is playing games. But underneath the box, the internals of an Xbox are essentially indistinguishable from any other computer. Indeed, an Xbox even runs a version of Microsoft's Windows operating system. It would be relatively simple for Microsoft to make an Xbox that could run virtually any game, not just games built for its own system. Even so, there are very few complaints about the fact that no one can play Sony's PlayStation games on Xboxes. So why should Microsoft complain that its own games cannot be played on an Apple device? How is an iOS device different from an Xbox? Indeed, in 2020 an estimated 72% of app store revenue across Apple and Google's app stores came from games, representing about $80 billion.[4] Given the money involved, it is easy to see why it is in Apple's interests to treat its device as a console. The questions remain. Is an iPhone a small computer, or is it a console? Or is it something else? Even if the device doesn't change, the choice of wording used to describe it has a major impact on how people react to it and what they expect from it. It even has legal implications—Epic Games sued Apple over its policies around the App Store. At stake in the lawsuit are issues such as how much control Apple can exert over third-party apps. These pragmatic discussions unveil the rhetorical function of metaphors that is used to emphasize some aspects of the technology (it does make calls like a phone), and hide or deemphasize others (but is also a general-purpose computational device).

Apple's claims that its app store is a "walled garden," an idyllic metaphor that suggests comfort and safety. This framing is good for Apple, as the associations of such a garden are almost uniformly positive. Apple's framing is one of protection; bad things cannot get into the garden to ruin the experience. Writing for the *Wall Street Journal*, Joanna Stern explored some of the implications of the metaphor and of Apple's rules that are not so idyllic, such as stifling innovation and keeping people locked into the garden.[5] Expanding on this, tech writer John Gruber suggested an alternative metaphor—a theme park. In this metaphor, Gruber compared Apple's store to a theme park where users are charged for admission but such parks are "fun, safe, and deliver a designed experience." He contrasted this metaphor with public parks, which have a different set of strengths and weaknesses.[6] Picking up on this, Dieter Bohn of *The Verge*, in an exploration of tech metaphors, suggested a completely different metaphor, that Apple is like a phone carrier.[7] Bohn's framing is much more negative, comparing Apple's policies to the very same phone companies that it has battled since the launch of the first iPhone. Such carriers are notable for forcing their own software onto devices, for example, exactly as Apple does. This discussion shows just how difficult it is to understand something as complex as the app store, but also how important metaphors are. Which of the three metaphors a casual reader happens to encounter can have a large impact on how they think about the App Store.

The ubiquity of CPUs and GPUs across many devices makes it hard to define and classify a wide range of objects. We have "smart" washing machines, refrigerators, and televisions. Most of these contain more computational power than the computers that first sent humans to the moon. With the introduction of the Series 6 Apple Watch, analyst Horace Dediu noted that the Apple Watch can now: sense blood oxygen levels, sense heart rates, generate an ECG, detect when sounds may be damaging to hearing, detect falls, and sometimes notify emergency services if the wearer appears to be unconscious, detect altitude, monitor hand washing, pay for groceries, and more. It can do all of these things because it has a computer inside, and yet none of the things listed are things that any desktop computer can do. It certainly can tell time, but is it a watch? Telling time is no longer a killer app when everyone has a smartphone in their pocket and many of the devices around our homes and offices display the time constantly.

As we noted in the introduction, some technologies exist mainly to enable other technologies that can in turn mediate our activities with the world. The examples in this chapter suggest that the chips that power smartphones and other devices are enablers that make the activities of smartphones possible but do not differentiate them from other technologies. Tablets especially make it difficult to see the uniqueness of smartphones. The obvious differentiator between tablets and smartphones is size, but smartphones continue to grow in size. Meanwhile, the differences between tablets and laptop computers also continue to blur, as laptops are increasingly enabled with touch UIs and tablets can be used with hardware keyboards and mice.

Similarly, it is tempting to argue that what differentiates mobile devices from other classes of technologies is that they are general purpose. Cory Doctorow has been writing and speaking about "the war on general purpose computing" for more than a decade.[8] In a nutshell, this war is happening because the more a device is capable of, the more harm it is capable of, and different stakeholders have very different definitions of harm. For example, digital objects can essentially be copied *ad infinitum*. In the early days of the internet, an application called Napster took advantage of this to allow people to freely share music. The entertainment industry saw this as an obvious threat and worked to shut Napster down and enforce Digital Rights Management (DRM) systems on media players. "Owning" a song or a movie does not mean that you can do anything you want with it; you are restricted by laws and regulations. As time has gone on, the number of these regulations and restrictions has continued to grow. The more that a device can do, after all, the more bad things that can also be done with it. As Doctorow noted "we don't know how to make a computer that can run all the programs we can compile except for whichever one pisses off a regulator, or disrupts a business model, or abets a criminal."[9]

So, inevitably, we return to predications and associations. A predication of smartphones that isn't shared by tablets is that they fit into one's pockets. Watches are not phones because they are worn on the wrist. When smart glasses make their inevitable return, they will belong on our heads. Thus, it seems that spatial predications, such as size and where they are used, are probably central to how we define and think about smartphones. This suggests that devices like iPods are probably most similar to smartphones because of their spatial characteristics. Meanwhile, the primary predication of smartphones that iPods have always lacked is that smartphones have a cellular connection. Finally, the touch UI discussed in chapter eight is a clear separator between what most people now think of as smartphones and the generation of phones that preceded them. In terms of forming a cognitive prototype, then, a good set of predications might be a device that fits into a pocket, uses touch UI, and has at least a CPU and a cellular connection. All of these predications are important, but one resonates most closely with the themes of this book, and it might be a surprising one—the fact that smartphones fit into our pockets such that they can be with us at all times.

With a smartphone, we have a useful device, one that is capable of mediating a huge range of activities and one that we have with us at all times. Together, this makes convenience a premium feature of the devices. Why carry a camera around when your smartphone comes with a quality camera? Why lug CDs around when your smartphone has access to nearly every song ever recorded? Why have newspapers delivered to your house when smartphones can show you the news literally as it happens? Further, these devices continue to grow in usefulness, adding apps, faster speeds, new sensors, and other features year after year. This means that our connection to them cannot help but grow as we use and rely on them for more and more things. According to App Annie, in 2021 the average user spent nearly five hours a day using their phone, the fifth year in a row where usage grew by about thirty minutes a day.[10] We use them for work and for leisure, to shop, to keep track of what we do, and to remind us of what is coming—the more activities they displace, the more they are intertwined with who we are. With such a device in our pockets, we can communicate with people across the globe, access a huge portion of human knowledge, and extend our normal capabilities as individuals in many other ways.

The Apple-Epic lawsuit also revealed, through a trove of Apple emails and testimony, a rather unsurprising fact about Apple's long-term strategy: Apple's goal is to get users as invested in its ecosystem as possible to keep them buying Apple products. An Apple Watch, for example, might work well on its own, but it works far better when paired with an iPhone. In the trial, it was revealed that Apple has not ported iMessage to Android simply because it would make it easier for users to migrate back and forth between devices. None of this is surprising. Apple is a business, after all. But the ultimate endgame of such a strategy is to essentially tie users to an ecosystem. You can send iMessages to people rather than specific devices precisely because of this approach. For now, iPhone users closely tie their identities to their phones, but Apple is trying to simultaneously strengthen that bond and widen it.

Other companies also want to strengthen their link to our identity. While Apple is still primarily a hardware company, many other tech companies create software. For a company like Facebook or Google, it is somewhat more challenging to link identity and their product because they do not control the device. Instead, they rely on tracking. Such companies have been shown to have a remarkable ability to keep track of our activities, mainly in order to build profiles of us to use in advertising. Thus a new battleground was forged when Apple added tools designed to limit a company's ability to do such tracking. This will drive software companies' need to find ways to identify users. It is no wonder that Facebook reportedly spent more than a billion dollars on the development of a watch by mid-2021.[11] TikTok, another example, announced in 2021 that it would begin to collect biometric data on users, e.g., if a user posts a video on TikTok the platform could analyze the video to determine who is in it, where they are, etc.[12] Software companies argue that such information will help them better serve users. Such an argument is a hermeneutic one, but it relies on the companies acting in good faith with regard to users. There has been concern about TikTok and data privacy, for example, since it is owned by a Chinese company, leading it to be temporarily banned by the Trump administration.

Throughout this book we have used the cognitive-hermeneutic cycle to examine how users and app developers act and react to one another in a continuous interpretive spiral. Such a spiral also plays out at the level of the device. The improvement in smartphones may seem small and incremental on a year-to-year basis, but over the span of several years, it is undeniably large. On the hardware side, processors get faster, screens get better and larger, batteries last longer, cameras take sharper pictures and video, and more. On the software side, some apps are improved and others are added with new functionalities. Meanwhile, users are driving many of these changes by showing their preferences with their buying choices. A 2019 survey showed that the top five consumer desires were: 1) more storage, 2) improved charging/battery, 3) more powerful processing, 4) increased durability, and 5) better cameras.[13] The first four can be summarized as "I want my phone to be always ready for anything" and the fifth as "especially as a camera." The importance of these five items can be seen in the ecosystems that have developed around them. The "cloud" is increasingly making on-device storage less important and is also available for processing. There is a large ecosystem of companies making mobile solutions for charging rundown batteries and another large ecosystem of cases and screen protectors designed to provide protection. In terms of cameras, there are portable tripods, "selfie sticks," and a host of other devices that can be used in conjunction with the camera; you can even turn your phone into a microscope with the right combination of hardware and software.

In the previous chapter, we argued that digital photos have helped to reshape our sense of self by mediating our memories, our communications, and our relationships. These are but a few, though clearly important, ways in which our devices are reshaping our sense of identity and our self-image. For many people the thought of being without their smartphone is terrifying because so much of what they do relies on that one device.

SPEAKING OF SMARTPHONES

The smartphone era began by piggybacking on the basic metaphor of telephones. The rapid evolution of its metaphorical spiral has been such that the device's use as a telephone is minimal compared to many other functions.

No obvious metaphor has sprung up to replace the telephone

Table 4: A partial, and growing, list of important aspects of self that can be mediated using one's smartphone

Aspect of Self	Mediation Examples
Communication	Text, email, phone
Relationships	Social media, dating apps
Memory	Photos, video, editing software
Purpose	Calendars, To Do apps
Health	Fitness tracking, fitness apps
Knowledge	Web browsers
Entertainment	Videos, web browsers, games, social media, readers

metaphor. Smartphones are like computers but are clearly a different beast. Similarly, they are also like tablets or even iPods. There are strong legal implications for whatever definition is settled upon.

Perhaps the defining characteristic of a smartphone is that it is always with us. Combining that feature with the myriad ways that they mediate our experiences, it might be said that the ultimate metaphor of the smartphone is that they are us.

The killer app of the modern smartphone is its generality.

PART IV

METAPHORS ARE CHANGING THE WORLD

This, the final part of the book, is divided into four chapters. The first three concern the fallout from the processes described thus far. These chapters are not comprehensive, as in many cases there is rich literature on the individual topics addressed, but they serve to provide a common framework that links the assorted issues. The first chapter deals with issues stemming from failures of metaphor. These failures could be because the chosen metaphor is not a natural fit for the technology, or because of cultural or ethical reasons. Such cultural and ethical issues give rise to the topic of Chapter 14, which examines who exactly is creating these technologies and metaphors. It will come as no surprise to most observers of technology that the answer is mainly white men living in one small corner of the United States. With that in mind, the following chapter deals with the ramifications of successful metaphors, metaphors that are able to engage users with the technology. Success, however, does not always mean that the results are positive. The final chapter is more forward-looking and offers some ways to take agency back from technology.

CHAPTER THIRTEEN

WHEN METAPHORS FAIL

COGNITIVE FAILURES

Despite all of the positive examples of metaphors working to be found in this book, metaphors are not a panacea. The cryptography community provides a cautionary tale of the kinds of things that can go wrong.[1] Indeed, there is growing evidence within that community that its chosen metaphors are actively harmful, framing critical issues in the wrong way and actually making it less likely that the technology will see widespread adaptation.[2] For example, the most basic metaphor in cryptography is probably the lock and key, where we are told that encrypting something is just like locking it up in a box and decrypting is equivalent to unlocking the box and gaining access to its contents. This metaphor may have worked as a description of older cryptographic systems. Unfortunately, it does not map well to most modern cryptographic systems, which instead are based on "public key encryption." In such systems users have both a public and a private key, and each key has very different functions. As Whitten and Tygar point out, "normal locks use the same key to lock and unlock, and the key metaphor will lead people to expect the same for encryption and decryption if it is not visually clarified in some way. Faulty intuition in this case may lead them to assume that they can always decrypt anything they have encrypted, an assumption which may have upsetting consequences."[3]

Thus, choosing proper metaphors can be a difficult task with significant long-term consequences. These consequences can be good or bad, depending on the perspective of who is judging them. Further, the initial choice may be sound, but as the technology is updated the metaphor may become a poor choice. And, as the cryptography example shows, the problems resulting from a poor metaphor may be difficult to overcome. On the other hand, as we saw with SMS, sometimes the cognitive-hermeneutic cycle can work to change the underlying metaphor over time. While the cryptography example may be just a case of the obvious metaphor not being the right one for the job, it does raise the specter that metaphors might be chosen to be deliberately misleading. Indeed, the cryptography community provides another example where a problematic metaphor has led to a kind of battle between different groups with different agendas. Jenner, in a blog post provocatively titled "Backdoor: How a Metaphor Turns into a Weapon," examines the multiple dimensions of this metaphor and the different predications that can be read into it. On one hand, governments, which want such backdoors, try to emphasize some of these predications, while people in IT, who warn against them, emphasize a different set, in what Jenner calls "a power struggle over its meaning."[4]

ETHICAL FAILURES

The world of cybersecurity affords another example of metaphorical failure, this time from an ethical perspective. In this world there are two kinds of hackers. Hackers with malicious intent are known as "black hat" hackers, while people that hack into systems in order to improve security are known as "white hat" hackers. These metaphors build on a constellation of metaphors where white represents purity and goodness, while black represents absence and the lack of morality.[5] We have noted throughout the book that there is a close relationship between metaphor and thought, with metaphor both reflecting how we think and framing how we take in information. Meanwhile, over a period of decades, a large number of researchers have been working with data collected from the Implicit Association Test (IAT), which tests a subject's associations by measuring reaction time.[6] While the IAT has been criticized as lacking the necessary accuracy for individuals, there is a great deal of evidence that it is accurate over large groups of people, and there are other confirming tests as well.[7] What the data collectively points to is a situation where the associations of whiteness with goodness and blackness with the lack of it transfer, even if only unconsciously, to racial attitudes. This should not be surprising given the literature on metaphoric transfer.[8]

We are not suggesting that the black hat/white hat metaphors are anything other than a reflection of wider cultural biases. That does not mean, however, that these metaphors should not be reevaluated and potentially replaced. Further, it does open the possibility that some metaphors could be created for less than ethical purposes, and for that and many other reasons we believe that it is important to examine the potential implications of any digital metaphor that gains widespread use. We outline some possible approaches to this task in chapter sixteen.

CULTURAL FAILURES

Developers of digital artifacts also need to consider how the meanings evoked by metaphors are different among their potential users. Technologies that increasingly target a global audience need to "translate" metaphors if their meanings lead to undesired interpretations within a given cultural sphere. For example, a particular flag can be considered a symbol of patriotism within one culture or country, but in another culture or country the same symbol might be considered abhorrent. Thus, a company with global reach might have problems deciding whether content including such a symbol fits their definition of acceptable. Further, a common object in one culture, such as a flag on a mailbox used to signify that a homeowner wants the mail carrier to pick up mail, may be completely absent in other cultures. Thus, a digital metaphor that includes such an icon may be lost on the intended audience. This issue is of growing concern when the world of smartphone technology is still dominated by white men working in Silicon Valley. If technology has an outsized impact on how we apprehend the world, then it should be concerning that these changes are guided by such a narrow group.

SPEAKING OF METAPHORICAL FAILURE

Metaphors can fail in multiple ways. They can be interpreted in ways that fundamentally misconstrue how a technology is meant to work. This can be the result of an unfortunate choice of an initial metaphor, poor ethical choices on the part of a developer, or differences between cultures.

CHAPTER FOURTEEN

WHO IS IN CHARGE?

Before we move on to the problems associated with metaphors working too well, it is worth looking at the group most responsible for both the successes and the failures of tech metaphors: the creators.

Since the creators of digital technologies have such a significant impact on the world today, as well as how we think about the world, they are worth examining. It is no secret that the majority of the companies building and shaping today's digital environment are headquartered in one place, Silicon Valley, and were all founded by white men. The stock market is dominated by the so-called FAANG stocks (Facebook, Apple, Amazon, Netflix, and Google).[1] Microsoft is not included because, despite being one of the five biggest companies in the world by market cap, it is not considered a "growth" stock. Tesla will almost certainly be added to this group. Every founder of these companies—Mark Zuckerberg, Steve Jobs, Steve Wozniak, Jeff Bezos, Reed Hastings, Larry Page, Sergei Brin, Bill Gates, and Elon Musk—is a white man, and white men dominate the work rolls throughout Silicon Valley. Only about 4 percent of professionals in Silicon Valley are Black, while about a third of companies have zero executives who are women of color.[2]

Thus, we have a situation where an unrepresentative group of people living in a tiny corner of the world have an outsized impact on nearly every aspect of modern life. Meanwhile, the architecture of the digital world gives those leaders, through the digital cognitive-hermeneutic cycle we are describing, an unprecedented ability to exert and maintain their power.

The urgency of this issue can be seen in the 2012 Facebook experiments aimed at directly manipulating users' emotions.[3] The scale of a company like Facebook is such that it can run such experiments in order to fine tune how users respond, essentially manipulating how users think and feel by choosing what they are shown. Since that experiment, Facebook has operationalized A/B testing, making it a core feature for advertisers. This is another example of how the normal hermeneutic cycle has been accelerated through digital technologies, and how companies are working to determine how they can optimize taking advantage of our thinking as effectively as possible.

Developers can also use these ideas in the opposite way. Sometimes developers and designers willfully do not attempt to bridge the semantic gap between the technical and the cultural, instead choosing to exploit it. This is generally done by making it difficult for users to do something that they may want to do. In the world of software such designs are known as "dark patterns."[4] For example, users may find it easy to sign up for a service that is initially free, but nearly impossible to cancel when fees begin, because the software has been carefully designed to make such details obscure. Designers could bridge the gap with a simple metaphor—an exit sign, for instance—but choose not to because helping the user does not suit a particular goal. In 2021 a consumer's rights group in Europe filed a lawsuit against Amazon for just such a practice.[5] Meanwhile, in 2021 the New York Times was criticized for writing an article decrying dark patterns even while its own website uses some of them.[6]

The level of control and the speed of change possible with digital technologies thus places unprecedented power in the hands of digital companies and the unrepresentative group of people that run them.

CHAPTER FIFTEEN

WHEN METAPHORS WORK TOO WELL

The argument that we have made in this book is that metaphors begin as a useful way to connect people to new technology, but due to the nature of metaphors and how we process them cognitively the learning process does not stop there. The metaphors continue to change how we categorize and therefore how we think. With regard to the technology that the metaphor serves, this can be a mainly positive outcome. Fifty years after a prototype device in a lab was coined a "mouse," the name has been fully incorporated into culture, even as the obvious links to the origin are being extinguished. There are many other cases, however, where metaphors have far more extensive ramifications. In these cases, the power of the metaphor may be such that the effects must be considered in any discussion of the technology's costs or benefits.

DIGITAL METAPHORS CAN DIVIDE US

We began this book with the idea that metaphors are an ideal way to communicate about what otherwise might be complex topics in relatively simple terms. Metaphors form a kind of cognitive short-cut that can allow us to build new categories using the foundations of existing categories. These categories are learned very quickly, accelerated by a potent combination of the metaphor, the pace at which technology improves, and the amount of time that people spend on their smartphones. One important effect of this process is that it can cleave the world into an increasing number of groups: those whose cognitive models have been altered by a particular metaphor and those who have not. This might be brushed off as "everybody" has a smartphone, or "everybody" is on Facebook, etc., but the truth is that many people do not have smartphones and even more people do not participate in any particular activity. Take Facebook, undoubtedly the world's biggest social media service. In 2020 Facebook was estimated to have just under three billion active users. We could try and discount how many of those users are corporations or bots, but even if we do not, Facebook's users make up less than 40 percent of the world's population. In North America, where Facebook usage is greatest, it is estimated that about 69 percent of the population is on Facebook.[1] Not everyone has been transformed by Facebook's model of friendship; the number of Facebook users is not particularly close to the world population.

Of course, all of this has happened before. The world has been transformed by technology many times in the past—the light bulb, the telephone . . .—the difference is the speed at which it happens in today's digital world. As we pointed out in chapter four when discussing the metaphorical spiral, TikTok reached two billion users in a matter of a few years, whereas even transformative technologies like light bulbs and telephones took decades to reach such high percentages of the population. During those decades, society had a chance to grapple with the effects that the technologies were having on the world. But it is also worth pointing out that, with those technologies too, society was broken into groups according to who had them and who did not.

As we have already seen, the differences in a word's meaning can have real consequences. Take the word "subscribe." In the predigital world, the kinds of things you subscribed to, such as newspapers or magazines, cost money. But the digital world has come to run on advertising, and so you can subscribe to things like podcasts and YouTube channels for free. On the other hand, there are many other services to which you can subscribe in the digital world that are decidedly not free. The result is that there are multiple meaning populations for the word subscribe. For some people it means paying money, for others it means that it doesn't, and for still others it might or it might not. Regardless, in 2021 Edison Research reported the results of a survey that showed that 47% of people who were familiar with podcasting assumed that subscribing to podcasts always cost money, leading to it being called "the stone in the shoe" of podcasting.[2] In 2021 Apple changed the language in its own podcast app to use the word "follow" instead of subscribe, prompting Podnews to describe it as "a small change with a big impact," because "the word 'subscribe' has been confusing potential listeners for more than fifteen years."[3]

The only way to effectively communicate with another person is to have shared signifiers. When the two authors of this book talk, we have an implicit agreement on most of the words we use, with occasional exceptions owing to the fact that one of us is a non-native English speaker. This agreement comes from the general overlap in our meaning horizons. When one of us uses a word like "justice," it is true that our personal experiences will shade the word's meaning in different ways, but it is also the case that we will understand what the other is conveying, and we will ask questions when we are not sure. Often our questions reflect the fact that we were raised in different cultures. To a large degree—and this is true for anyone—our ability to communicate successfully stems from the fact that we inhabit the same world and the objects that make up that world are the same. If one of us had poor vision or were colorblind, we might see trees differently than the other, but we would still agree on what the word "tree" means. With the advent of digital technology, however, it is now easily possible to at least partially inhabit virtual worlds that others do not, or to inhabit parts of the world that were previously inaccessible. While these things do afford many exciting opportunities and experiences, they can also come with a cost.

Sharpening Existing Divisions

Another important consequence has to do with the manner in which digital metaphors and algorithms have been implemented. By now, it has been well established that the digital world has problems with race,[4] gender,[5] class,[6] and age.[7] This is not surprising. As we have demonstrated throughout this book, technologists have worked hard to mimic the "real" world in creating the digital world, and the real world is full of racism, sexism, etc. Algorithms are not any sort of protection from these problems. They often only make them worse, especially since algorithms are hailed as being objective and thus supposedly immune to such issues.[8] Algorithms have also struggled greatly to mitigate problematic online speech.[9] One reason for this struggle is that in the cognitive-hermeneutic loop users, especially large groups of users, do have agency. For example, anti-vax groups eager to avoid Facebook's crackdown on vaccine misinformation during the coronavirus pandemic communicated using coded phrases designed to outwit Facebook's algorithms. So, the groups refer to themselves by names such as "dance party" and to vaccinated people by terms such as "swimmers."[10]

Meanwhile, algorithms can become racist (or sexist, or, . . .) in at least three ways. First, they may not actually be objective at all; they may simply codify privilege. Second, any algorithm is a model of a process, and if the process itself is racist, then it simply implements that racist system. In computer science this principle is known as "garbage in, garbage out," reflecting the fact that any algorithm is limited by the quality of its data. Finally, algorithms that learn will begin to reflect their environment. There have been several notable attempts to embed learning algorithms in the internet that have quickly learned to become racist, sexist trolls.[11] These are all serious problems for the digital world. They are made far worse by the hermeneutic cycles we describe in this book. As we have shown, the digital environment is optimized for learning—communications are simple and repetitive and many people spend much of their lives on their devices. As Sara Wachter-Boettcher explains, "the more technology becomes embedded in all aspects of life, the more it matters whether that technology is biased, alienating, or harmful."[12] The result is that "the new Jim Code" as Ruha Benjamin puts it,[13] is being widely assimilated everywhere, even by young children. Even worse, it is being exported around the world.

Culture as Digital Group

As we saw in Part III, the power of smartphones is that they remove barriers of time and space. At one point in history, you could only talk to another person if you were in the same place at the same time, but now you can have a conversation with someone regardless of location, and by using text, the conversation does not even need to be synchronous. Similarly, for much of human history the groups to which we belonged—our culture—were defined mainly by a combination of time and place. This is no longer the case; we are free to choose our groups, and the digital world offers more choices than ever.

Once upon a time, a role of culture was to provide individuals with the wisdom of the group, thus scaling and improving individual learning. For example, a religious admonition not to eat a certain food might be grounded in hard-won lessons about the dangers of said food. At its best, digital culture can play the same role. Users can find lessons to do virtually anything on YouTube, or learn math through online learning systems like Khan Academy. This is the promise of the digital age, that everyone has access to collective knowledge and that we can connect across time and space. There are, however, downsides to this freedom. Among them, not all online knowledge is hard-won or even real. We are replacing culture, with its emphasis on collective wisdom tested by time, with more ephemeral things, some of which are not well connected to reality. Disinformation and propaganda are nothing new in the world, but they too gain power from the amplifying effects of digital metaphors and the reach of the internet.

Digital Connects People and Distances Them

The fact that we are focusing on the ways that digital metaphors increase divisions runs contrary to the great promise of the digital world, which is that it can connect us regardless of distance. The coronavirus pandemic revealed both the truth and limitations of this promise. Digital metaphors are removing barriers of space and time to concepts such as conversations and meetings, but they are simultaneously removing other aspects of those concepts, notably those related to physical presence. A conversation over Zoom instead of text may restore some physical elements of conversation in that we can at least see each other's faces, but it also places unusual demands on our attention and does not capture body language in ways that in-person conversations do. The uncomfortable feeling that this engenders are so widespread that it already had a name, "Zoom fatigue," and academic papers were written about it before the pandemic was even over.[14] As a species the vast majority of our evolution centered around communicating with each other in close proximity, and with learning and using many physical cues to do so. Communicating remotely is not the same and it never will be, no matter how much we do it. Digital connection is a substitute for physical connection, but it is far from a perfect substitute.

ALGORITHMIC ACCELERATION

Throughout this book we have shown that the combination of metaphors and digital technology is a powerful learning accelerant that can drastically speed up the cognitive-hermeneutic cycle. Digital technology can make this combination even more powerful through the use of algorithms. While tech developers can iterate quickly compared to older technologies, they are still limited by the speed at which humans can work. Algorithms do not have such limitations. They can respond to users immediately, essentially supercharging the cognitive-hermeneutic cycle to its maximum speed. The potential effects of this can be seen in the riots at the US Capitol in January 2021.

On January 6, 2021, the United States Capitol was stormed by angry insurgents. As we write this book, there are still uncertainties about what exactly happened and who was responsible. Some of this uncertainty has been created by groups looking to deflect blame. As we previously stated, digital groups such as QAnon have replaced cultural ones for many people. By January 14, the *New York Times* published an article examining the role of social media in the mob's behavior.[15] The article does not use the language of this book, of course, but it is a nearly perfect match for the cognitive-hermeneutic spiral and its results that we describe. The article traces the online life, or more precisely the Facebook life, of a pair of individuals, seemingly normal people posting pictures of family and friends and generally living up to Facebook's intrinsic metaphors. Their cognitive-hermeneutic spiral took a turn soon after the presidential election in November 2020, when they made posts that generated an unusual number of likes and comments. Thus began an accelerated spiral. From Facebook's point of view, posts that drive engagement are a solution to the problem of holding users' attention; the algorithms suggest such posts to other users, further driving likes and comments. In the meantime, Facebook is also algorithmically analyzing the posts and using them to suggest articles and groups to the poster. All the while the person who made the original post is also getting feedback—likes and comments–which are the reward structure of social networks. The role of rewards in learning is probably the oldest area of study in the field, going back to Pavlov's dogs and continuing today in machine learning. When presented with an environment that contains rewards, learners will optimize their behavior to maximize those rewards. In 2021 researchers at Yale, looking at posts expressing "moral outrage" on Twitter, found exactly that pattern.[16] For the individuals in the *New York Times* story, that means optimizing behavior with respect to the kind of posts that drive likes and comments. Facebook is happy to play along by getting like-minded people together in groups and recommending similar content. This kind of behavior is as old as the internet, going back to when early internet companies learned how to optimize their websites to take advantage of Google's original "page rank" algorithm. In a system like this, the accelerants are so powerful that seemingly "normal" people can be radicalized over a period of only two months.

The other side of the learning spiral are the companies that model user behavior in order to keep them engaged. In 2021, the *Wall Street Journal* ran a powerful experiment that showed how well this works.[17] They programmed some bots—artificial online personas—with specific interests and then signed the bots up with accounts on TikTok. They found that TikTok's algorithm was able to quickly identify what the bots liked. It wasn't even necessary for them to use likes or hashtags to do so; just the amount of time spent looking at different videos was enough signal for the algorithm to highly tailor the bots' feeds and send them down specialized "rabbit holes." The authors found that if they programmed a bot to prefer, for example, depressive content, within a matter of hours the TikTok algorithm would tailor as much as 93% of the bot's feed to that content. Why not 100%? As we previously discussed when looking at learning, learning systems have to trade off exploitation, in this case showing another video that the user will definitely like, versus exploration, showing videos from other genres such that other genres might be discovered. A given user might grow tired of a genre after a while, after all.

DIRECTED ATTENTION FATIGUE

As we noted when discussing the metaphorical spiral, one of the things happening with digital metaphors is that they are all competing for our attention—to get it, and then to keep it. Over time this competition invariably leads to better and more effective ways for apps to keep us engaged. This is reflected in usage metrics for digital devices. There is a cost beyond the time involved, and that cost is to our attention. There is a growing and robust literature which we reviewed in chapter six that details how directed attention is a limited resource that can be overused and fatigued.[18] The result of fatiguing this system is the loss of "executive function" abilities, which has been tied to diminished cognitive functioning, decision-making, and health, among other factors. A common reaction to evidence of attentional fatigue is to suppose that activities like watching funny videos online might actually help relieve it. Berman and Kaplan, however, argued that television, a reasonable proxy for this kind of activity, "is a counterproductive means of restoring directed attention."[19] This was more recently supported by Basu and colleagues, who noted that restorative activities typically allow the mind to wander.[20]

The Zoom fatigue phenomenon is yet more evidence that spending time online only depletes attention. The kinds of activities that effectively relieve attention fatigue typically involve natural settings, walking, and being disconnected from technology. Thus, the addictive nature of our digital devices makes for a kind of vicious cycle where we use them too much, which in turn leads to cognitive fatigue, which in turn impairs our decision-making and makes us more eager to engage in activities that feel cognitively easy, which in turn leads to us spending more time connected. A common complaint of the modern world is that we are losing our ability to pay attention. It is actually more likely that we are simply keeping our attention system in a constant state of exhaustion.

SPEAKING OF METAPHORICAL SUCCESS

What is good for an individual technology may not be good for individual people, or even for society as a whole. The amplifying power of technology is such that many tech metaphors simultaneously build new groups and make the resulting groups more disconnected.

One way that the marriage of technology and metaphor has strongly impacted individuals is on our attention. As apps get better and better at grabbing and holding our attention, our ability to deploy our directed attention to achieve our goals is weakened.

CHAPTER SIXTEEN

METAPHORICAL CRITICISM

The ambiguity of metaphors is both a blessing and a curse. On one hand, it enables new possibilities of expression, but on the other it opens the door to manipulation and disguise through indirect associations or simply through unwanted interpretations. Wyatt, among others, has repeatedly argued that metaphors have not only a cognitive but also a normative dimension. She called for the careful examination and criticism of how metaphors are used in technology, and in many ways we see our book as an answer to her call to arms.[1]

Blavin and Cohen also highlighted the implications of metaphors in the ways legislators deliberate on new topics, particularly related to new technologies. They insightfully summarized their view on metaphors' intrinsic ambivalence, saying that "while metaphors aid humans in comprehending abstract concepts and legal doctrines, they also may limit human understanding by selectively highlighting various aspects of an issue while suppressing and marginalizing others."[2]

We suggest, following Ricoeur, that an examination of metaphors requires two complementary and interpretative movements.[3] The first movement should be one of openness, embracing the new possibilities suggested by metaphors. Such a movement can emphasize and explore the positive characteristics that metaphor has in unveiling new experiences and phenomena through innovative predications capable of disrupting the otherwise solidified canon of lexical expressions. The fundamental premise of this first movement is based on the ubiquity of the metaphorical phenomenon which, as we have explored extensively, is seminally and intricately linked to the malleable and comprehensive nature of natural languages.

Robert Frost said that:

Unless you have had your proper poetical education in the metaphor, you are not safe anywhere. Because you are not at ease with figurative values: you don't know the metaphor in its strength and its weakness. You don't know how far you may expect to ride it and when it may break down with you. You are not safe with science; you are not safe in history.[4]

In this talk, delivered at Amherst College in 1931, the New England poet invited the academic community to look to metaphors as a fundamental mechanism for understanding meaning—not just for poetic discourses, but also scientific and historical discourses. Frost further suggests that metaphor is the whole of thinking, and, therefore, learning about metaphors is learning about thinking.[5] Two decades, later this idea found new life in the work of Lakoff and Johnson. We would also gladly add technological discourses to Frost's list—delivered in words, symbols, and digital artifacts.

Nevertheless, this first movement of openness needs to be balanced by a second, critical one, in which the ambiguities of metaphors are explored so that their shortfalls and possibly misleading meanings can be unveiled, criticized, and, in some cases, denounced and combatted. Metaphors can be risky and manipulative. It is necessary to recognize that there is a breaking point in metaphorical statements, at which they cease to be productive and instead become mechanisms of illusion and manipulation. Metaphors always both show and hide aspects of the reality they describe. And it is often necessary to unveil not only what is innovatively said by the metaphor, but also what is left out, and sometimes intentionally concealed.

In *Leviathan*, Hobbes, imbued with the spirit of the European Enlightenment, railed against the ambiguity of metaphorical statements and proposed an alternate ideal borne of precise language, free from ambiguities. Thus, he ironically rejected metaphors using the *ignes fatui* metaphor. He suggested that, like flashes of light that occur in marshlands and bogs—*ignes fatui*—which produce only evanescent light sources but provide decomposing organic matter combustion, metaphors are not good guides for rational thought.[6]

More recently, several scholars have devoted their attention to the manipulative use of metaphors aimed at the maintenance of power.[7] For example, Andrew Goatly has done an extensive study of how metaphorical themes such as "important is big," "status is high/achievement," "good is pure/white," and "time is money" are used in ideological discourses to justify structures of power and domination.[8] More directly connected to this book, the Alexander Humboldt Institute for Internet and Society organized a series of essays by a variety of authors under the general heading "how metaphors shape the digital society."[9] Each post examines different tech metaphors and how the choice of metaphor has impacted society.

Nevertheless, the centuries that separate us from the modern goals of unambiguous language dominated exclusively by scientific rationality have taught us that human thought is intrinsically metaphorical. Purging metaphor is purging the creative and imaginative potential that drives new discoveries, opens up new perspectives, and keeps us individually and collectively alive. There is no non-metaphor place. Nevertheless, Hobbes's caution should resonate in our minds as an alarm that warns of the constant and imminent danger that comes with the use of natural language and the way we think metaphorically. This is also true for the analysis of metaphors used and implemented in digital artifacts.

In light of this mandate for a critical interpretation of digital metaphors, we reexamine, through a critical interpretive lens, two widespread digital metaphors related to social interactions and digital data.

To structure this metaphorical critique, we should recall some key concepts that help us understand particular traits of metaphors. In the relationship between the *source/vehicle* of a metaphor and its *target/tenor*, there is a process of mapping predications and cognitive associations based on similarities. Such mappings are delineated by the *context of the use* of the metaphor that we refer to as the *ground*. Thus, some, but not all, associations and predications relevant to the source are applied to the target. In this process, an unexpected, impertinent predication is created within the scope of established lexical rules suggesting a possible new meaning. However, not all typical source predications have the same emphasis in terms of the metaphor. The context of the metaphorical statement highlights some source predications, we will call these *explicit*, but it leaves out others that we will call *latent*. Similarly, metaphors will call to mind some of the target's characteristics and associations while leaving others untouched by the predication. We will call the former *explored target features* in contrast to the *unexplored target features* within the grounds established by the metaphorical statement.

Finally, there is a lot to say about the relationship between metaphors, but we will highlight just two of them proposed by Goatly. The first relationship is *diversification* (see Figure 10), in which the same target is referred to by multiple sources. The second is *multivalency* (see Figure 11), where the same source refers to multiple targets.[10]

Before using these concepts to describe some strategies to criticize digital metaphors, a literary example from Shakespeare's

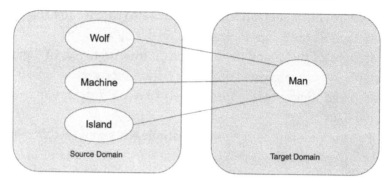

Figure 10. The same object can be targeted by many different sources. E.g. "man is a wolf," "man is a machine," and "man is an island."

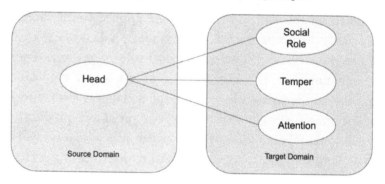

Figure 11. In multivalent metaphors the same source could be used for a number of different targets. E.g. "she is the head of the department," "they are hot headed," and "keep your head in it!"

monologue "All the World's a Stage" in the pastoral comedy *As You Like It* will hopefully clarify the conceptual framework a little more.

> All the world's a stage
> And all the men and women merely players;
> They have their exits and their entrances,
> And one man in his time plays many parts.[11]

In this excerpt, one of the metaphors asks the reader to see the *target/ tenor* "men and women'" as the *source/vehicle* "players." The context of the metaphor is provided by the comedy that delimits the *grounds* of the metaphor, encompassing changes in the lifespan of women and men. The *explicit source predications* in this segment are exits and entrances (from the stage), and to play many parts. Some *latent source predications*—ones not suggested in the metaphorical statement—are, for instance, players use customs and players receive a salary.

Regarding the metaphorical target, some *explored target features* are the aging process and the different social roles men and women take in their lives. Among the multitude of *unexplored target features*, we could highlight the fact that men and women may get sick and die prematurely. Through the concept of *diversification*, we could consider other *sources* to the same *target*, for instance, a marionette that would suggest the metaphor "men and women are marionettes," from which we could explore other *target features*. Finally, the concept of *diversification* leads us to consider other *targets* for "players," such as the employees of a corporation, which would unveil other possible predications of players not explored by the initial metaphor.

Equipped with these concepts about metaphors, we suggest some basic strategies that could be applied when critiquing digital metaphors.

(1) Clearly identifying the metaphor in the form of *seeing X as Y* is an important initial step. It requires the clear definition of source and target, as they could be confusing or not explicitly mentioned when the metaphor is proposed by digital technology. Because conventional metaphors can become transparent due to their constant use, the simple identification of an underlying metaphor may lead to important insights and critical considerations about the digital technology—am I really talking to someone when typing characters on the keyboard of my phone?

(2) Next in the flow proposed by the critical analysis is an exploration of metaphors' heuristic and innovative potential. At this initial interpretative moment, a detailed description of *explicit source predications* sheds light on the scope of the metaphor, and a refined analysis of the *explored target features* clarifies the aspects of the phenomenon captured by the metaphor and how the impertinent predication provides cognitive access to such elements.

(3) In the subsequent analytic moment, we take a more critical approach and explore limitations, shortfalls, and risks of the proposed metaphor. At this stage (3a), it is imperative to discuss the *latent source predications* that are not explicitly explored in the metaphorical context but that can influence and distort the understanding and valuation of the underlying phenomenon mediated by the metaphor. They point to cognitive and emotional associations of the metaphor that are only subliminally suggested but have relevant effects on the values and actions shaped by the metaphor. (3b) It is also crucial to investigate *unexplored target features*, because they can unveil positive or problematic characteristics of digital technologies that are not adequately captured by the metaphorical statement. (3c) The analysis of other metaphorical interactions through the concepts of *multivalency* and *diversification* allows for the exploration of other associations evoked by the source and target and identification of additional *unexplored target features*. It is also possible to propose alternate sources to capture the nuances of the target, and the conflict between the interpretation proposed by different metaphors for the same digital technology broadens the understanding of the underlying phenomenon.

This interpretive method should offer a deeper and more balanced perspective of the cognitive gains and possible manipulation risks arising from employing metaphors in digital technologies. At the very least, this analysis should promote a discussion of the implications of using certain metaphors and eventually foster the replacement or removal of inadequate and misleading metaphors in which poor or ill-disposed associations cloud heuristic, innovative, and cognitive gains. In the following two sections, we apply this method to the analysis of the Facebook like metaphor and some metaphors related to digital data.

THE LIKE BUTTON

We start with the seemingly naïve and harmless metaphor of the thumbs-up button to express appreciation for a post on Facebook. The famous like icon had an exclusive reign on the app's interface from 2010 until 2017, when other "reactions" were added, though with reduced visibility in comparison to the like. The explicit metaphorical content of the like button is relatively straightforward and allows us to express feelings of approval for ideas and information shared by our Facebook "friends." In this sense, as we mentioned previously, this metaphor was an effective way to implement a feature designed to promote positive engagement between people. From Facebook's perspective the positive affect associated with this is an added benefit, as that affect will naturally transfer to the metaphor and to Facebook itself.

However, in the second critical step of this analysis, we should consider the other associations and possible motivations behind Facebook's choice of this specific metaphor. If we think about the target domain of this metaphorical theme—the friendship relationship—we must face the fact that friends express more than just positive feelings. Friends are often the people we turn to when we need honest feedback. One of the possible underlying reasons why Facebook has ignored this is linked to the company's fundamental business agenda: to increase the number of users and their engagement. Thus a positive relationship environment tends to increase user satisfaction with interactions within the application and by association with the app itself. But still other layers of intentionality behind the metaphor can—and should—be explored. In September 2015, *MIT Technology Review* published Facebook's plan that the like and share buttons that publishers have added to their pages and mobile apps would start funneling data on people's web browsing habits into its ad targeting systems.[12] Furthermore, with machine learning-empowered recommender systems, every like in a post (even noncommercial ones) becomes a signal of user preferences that is used for profiling and targeted advertising. These user models can also help to fine tune the feeds that help keep users engaged. Another problematic dimension behind the like metaphor and similar ones is the concern, particularly acute in adolescents, of a culture that emphasizes dependence on positive feedback from friends in social networks, with significant psychological and behavioral repercussions.[13]

Fundamentally, the like metaphor is part of the metaphorical bundle for "friendship" proposed and implemented by Facebook. Indiscriminate likes and endless lists of "friends" flatten the very concept of friendship, affecting and redescribing one of the most fundamental relationships in contemporary societies.[14] The metaphorical critique proposed here should lead us, personally and collectively, to reflect on the extent to which we want to embrace this new concept of friendship. The metaphor of friends that was originally used to make and explain computational concepts now works in the opposite direction, changing the very meaning of friendship, friends, appreciation, chatting, etc. Digital metaphors are full of meanings that affect how we see and value our experiences, relationships, and ourselves.

It is possible to take another step and think of other metaphors that can be and are implemented in social networks to evoke and incite other relationships. As always, these new metaphors will also have their limitations, and they also need to be criticized and put into dialogue with other metaphors. Through the vast set of meanings revealed by this conflict of interpretations, new ideas and possibilities can arise for digital technologies to use metaphors more conducive to inspiring ethical and social values.

An example of a distinct metaphorical theme for social networking applications can be found in the Junto application. According to its designers, Junto intends to "inspire authenticity and rebalance our relationship with technology."[15] One of the app's explicit strategies to achieve this goal is selecting new digital metaphors to reframe the social networking experience. First, it is interesting to remark that the name of the application itself is a metaphor expressed in Spanish and Portuguese without a lexically similar term in English, which changes the interpretive bias of the metaphor for speakers of other languages and opposes the centrality of the English language in technological metaphors as mentioned in the previous sections.[16]

Junto designers also propose a set of innovative metaphors for its essential features. The metaphor "expression" replaces the metaphor of "post," and there is a public virtual space for exchanging expressions with everyone called "collective." Another fundamental metaphor the app implements is "packs." According to the designers, a pack is "a group of close friends (and/or family) who evoke the most unfiltered and honest version of yourself and with whom you can share, privately."[17] Surely, it is a metaphor that rouses other cognitive and emotional associations whose meanings must be explored and contrasted to its more commonly used alternatives. While "pack" certainly evokes some predications and associations intended by the designers' mission statement, such as familiarity and common goals (Cub Scout Pack), it can also evoke associations of competition, ferocity (pack of wolves), and even illegal activities (a pack of thieves).

The forms of interaction that such semantic changes can provoke in user communities is another fundamental aspect of the critical analysis of this metaphor. An essential change proposed by the metaphorical theme and the underlying intentionality of Junto is the removal of the like metaphor. In the current version of the application, you can only interact with expressions in textual comments, which certainly changes the dynamics of content-mediated relationships and avoids some of the potential negative effects of the like button.

As of this writing, Junto is still in development on Google Play and has a small community of early access users. However, what interests us most is the example of digital artifacts that propose to structure their functionalities and interfaces through different metaphors when recognizing the limitations and practical implications of existing metaphors.

STREAMING FROM CLOUDS OF GOLD

In January 2015, Tim Hwang and Karen Levy wrote a short but very thought-provoking article in *The Atlantic* about the importance of the metaphors used to describe digital data.[18] Recognizing the difficulty of achieving a precise and clear definition of the concept of digital data, they highlighted three metaphors commonly used to express how we relate to digital data: "data streaming," "data mining," and "data cloud." It is fascinating to reflect upon the different materialities suggested by these three metaphors— and their cognitive and emotional implications. Data can be fluid like a stream of water, it can be solid and impenetrable enough to require mining efforts, or it can be so gaseous that it evaporates and condenses only in the form of clouds that we contemplate in the distance.

Following our metaphorical criticism strategy, let's start our analysis with a movement of openness and receptivity to explore new meanings—typically with positive connotations—that are usually publicly emphasized, and expand and help us understand relevant aspects of digital data. In the second movement, we complement this first approach by exploring the inadequate, inefficient, or elusive latent associations of these same metaphors. And finally, we explore alternative metaphors and interpretative critique that should be considered when interacting with digital metaphors.

As discussed in previous sections, the "data is a stream" metaphor was a useful way to indicate the new features and conditions of digital media consumption that were no longer accessed locally via magnetic tapes or CDs but were continually accessed from remote storage locations. Streaming data indicated the need for users to connect with the spring/repository from which the data originates; it also provided the implication that the data flow can sometimes be interrupted or become insufficient to properly consume the media. Streaming video and streaming music have become common metaphors and continue to productively shape how we interact with data.

Data mining has been a buzzword for computer science over the past two decades. Graduate and specialization courses spread, consultancies offered data mining service as a differentiator, and new software platforms were created specifically to optimize mining processes. It effectively created a "digital gold rush" that fully justifies the metaphorical theme suggested by mining. The metaphor suggests an active process of expending great effort to extract something that, when found, can have immense economic value— "data is gold." The metaphor has been fully adopted in some activities, such as advertising. Advertisers use the patterns in mined data to optimize the return on their investment by reaching the potential consumers who are most likely to pay for their products. Such customers are the "gold" in the data for advertisers. In other application domains, such as diagnostic imaging, the result of data mining to identify diseases early and effectively also reinforces the idea of the value that can be extracted from the process of inspecting a large volume of data.[19]

The data cloud metaphor is the most open—and therefore potentially the most problematic—of this triad of metaphors. It suggests that the data floats freely in the internet's skies, so you don't have to worry about storing it in a local container. Using one of the metaphorical themes linked to directionality, you can upload your data to the cloud, and this brings with it the connection that "up is better." The cloud offers yet another way that the digital world disconnects us from constraints of place.

The evident weakness of this metaphor suggests that we begin our second moment of metaphorical critique of these three metaphors in reverse order. The data cloud metaphor is extremely poor with regard to its potential associations with the phenomenon of data storage in remote locations. It is so vague and limited that it fails to inform users' understanding of what is happening, creating an almost mystical concept around data storage and completely hiding some crucial implications of the process. It is a clear case where what the metaphor hides is far more important than what it reveals and communicates. Consequently, a significant issue with using the cloud revolves around trust. How can you trust something that you do not understand?

Regarding the vehicle's predominant associations, such as transience, mobility, and intangibility, none seem to clarify the tenor linked to massive computational structures that occupy vast spaces and need a lot of maintenance due to their physical nature. There is nothing gaseous or transient in the data centers and magnetic storage devices used for data clouds. Other aspects of the tenor (physical data storage infrastructure) are obscured by the metaphor, such as the amount of energy that data center computers consume and its potential impact on the environment.[20] In addition, the "foggy" idea of the cloud also hides the reality that all data is actually being stored and becomes the property of companies that can use this information for various activities without users' knowledge. It also does not address the numerous security issues inherent in having your data stored by third-party systems. This very brief critical analysis of the data cloud metaphor raises a number of problematic consequences of the use of a weak metaphor, a metaphor that hides more than it illuminates and thus becomes almost meaningless. Yet the data cloud metaphor refers to a fundamental phenomenon of digital technologies, and its poor construction only distracts users from problematic aspects of the reality.

Let's now make a brief critical analysis of data mining, a metaphor that clearly has more significant heuristic potential than data cloud, but which still needs to be explored so that other understated aspects of the proposed associations are properly considered. As we have seen, the idea of mining has a strong connotation with the production of economic value, making the positive association of productivity lead to considering investment in the activity as something evidently profitable. However, a more nuanced analysis of the processes of extracting value from large volumes of data reveals their limitations and the uncertainty that gold will actually be found through this mining.[21] Another extremely unsettling aspect of this metaphor is that mining is typically done on natural resources that are either public or owned by the mining entities. In data mining, the field being mined is often the data of users who are unaware that they are "donating" their resources for this type of commercial exploitation. Further, "data as gold" suggests that individual chunks of data are what is worthwhile. This is fundamentally backwards; it has been suggested that "data is sand" would be a better metaphor.[22] Any individual piece of data is fairly useless, it is only when it exists in large quantities that there is actual value.

Finally, we take a critical look at "data is a stream." One of the things this metaphor leaves out is that there are different channels for the data stream, and each of them has different financial implications for those who decide to drink from this digital source. Anyone who had a bad surprise with their mobile phone data plan bill because they watched movies outside of Wi-Fi coverage will have painfully learned that the particular channels have significant differences, differences not directly captured by the metaphor despite the spring and liquid being the same. Another aspect of this metaphor that is not usually communicated is that the owners of the data stream pipelines can, depending on the laws of the country/region, control the data flow or even prevent a specific type of liquid from passing through the stream. This little-explored part of the metaphor is behind the vast legal controversy in the United States over net neutrality. Unlike pure water streams, data streams are made up of different "liquids"—the types of data that travel, for example, whether the data is from a Netflix movie, a page from your favorite newspaper, or a song from Spotify. These diverse liquids can be easily filtered by whoever controls the data stream channels. Thus, the movie stream can be removed or have its flow significantly diminished, while the stream of information pages continues with the same flow of data.

This last analysis illustrates how it is possible to enter the metaphorical theme. This can be done to clarify other dimensions of a complex technological issue and foster other political and ethical arguments. Such discussions can be more nuanced and multifaceted, providing interpretations accessible to communities with varying degrees of technical expertise.

SPEAKING OF METAPHORICAL CRITICISM

The unrivaled combination of the power and flexibility of metaphors in conjunction with their use in technology requires careful monitoring. As a compact way of summarizing things that are often very complex, metaphors run the risk of being misused and misunderstood.

Important tech metaphors should be pulled apart and examined, both in their construction and in their intent, with an eye for what is implied and what is obscured or hidden.

REVIEW AND CONCLUSION

Now that we have concluded the main arc of the book, it is time to take stock and revisit some key ideas. For starters, we want to highlight the title of the book and its relationship, through our proposed cognitive-hermeneutic method, to the impact of digital technologies. When we first came up with the title *Meaningful Technologies*, we had four intertwined ideas in mind, each concerning an in-depth analysis of digital technologies.

First, we emphasize that digital technologies are more than neutral tools whose impact is defined exclusively, or even mainly, by how they are used. They are full of meanings, and throughout the text we have proposed that digital metaphors are at the heart of these meanings, embedded and implemented in digital technologies that affect users and society.

Second, we highlight, through a conceptual framework based on cognitive science and hermeneutic philosophy, that an analysis of the ways in which digital technologies affect the meanings we attribute to our relationships with the world and with others reveals a deeper aspect of their impact on our lives. By reshaping our semantic horizons through digital metaphors, they alter the cognitive foundations that form the fabric of our personal and collective identities. Technologies make sense; they engender new senses and meanings into our individual and societal structures.

Third, we point out that, precisely because they are full of meaning and because they transform meanings, digital technologies are important—meaningful—and must be analyzed and better understood in terms of their complexity and influence.

Finally, as we suggested in the last chapter, analyzing how digital metaphors mediate the relationship between technology and meaning not only enables us to describe and criticize them, but also gives us the potential to propose changes capable of altering this fundamental and transformative layer. To make sense of technologies is to bring sense into technologies by transforming how they embed meanings—and digital metaphors are a crucial mechanism through which we reshape digital technologies' proposed meanings.

Our title, and thereafter the book's trajectory, was thus an invitation to rethink—through digital metaphors—our collective responsibility to redefine digital technologies to provide a kind of road map such that they can become carriers of constructive and beneficial meanings for contemporary society that might better promote the common good.

We also want to call back to the foundational ability that makes metaphors so prominent and useful in language—seeing one thing as being something else—and pair it with the old adage that "seeing is believing." Digital metaphors take this one step further. When Amazon tells us that an icon on our screen is a shopping cart, they literally make the pixels resemble one. It takes a real act of will, then, to remind oneself that there is no shopping cart, or that friendship on Facebook is not the same as friendship in real life, or even that Google's selection of photos as a memory of my vacation might not actually be representative of what happened. Whether that metaphorical shopping cart, friendship, or memory is better than what preceded it is up to the individual, but it is a consideration that needs to happen more often than it does now, and it needs to be considered in a larger context too. Just as we cannot truly understand a sentence without knowing its context, so too we cannot truly understand a technology or its metaphor without connecting it back to the world.

The book is also an invitation to consider the intricate symbiotic relationship between technology and society through the lens of our proposed cognitive-hermeneutic spiral of digital metaphors. On one hand, digital metaphors are born of designers' need to create bridges so that their technological enablers will be understood, typically by relying on ideas and other technologies that are already well established in society. On the other hand, the massive and swift adoption of digital technologies and their subsequent repetitive use speed up the transition of digital metaphors into the linguistic lexicon, restructuring how we talk and, in turn, how we think about social aspects of our lives, ranging from things as simple as getting a ride to the more profound, such as the presidential election process. And once they are assimilated, digital metaphors give birth to something new, truly transforming society by fostering the creation of new realities that will, in turn, be the basis for new technologies and subsequent digital metaphors.

Our realization of how central digital metaphors are in the interplay between society and technology inexorably led us to ethics, culminating in the proposal of a new kind of metaphorical criticism that we presented in the previous chapter. Recent discussions about the harmful impacts of social networks on individuals and societies have highlighted the need for ethical debates on new technologies. The speed and scope of digital technologies' impact—which only continues to grow due to increasingly robust and sophisticated architectures—requires that the effects of new technologies be analyzed and criticized even before they are widely deployed. As we have discussed throughout this book, we believe that the impact of new technologies cannot be fully understood without a thorough examination of the digital metaphors that they rely upon. It is necessary, for example, to investigate which ideas and values inherent in a technology are emphasized on one hand, or obscured on the other, by the particular digital metaphors selected. In this we are aligned with Wyatt, who called for critical scholars "to be simultaneously careful and imaginative in their own choice of metaphorical language."[1] We would expand the list of those who need to be careful to developers and even to users. The goal would then be to propose alternatives so that their impact can be better aligned with the common good. We believe the cognitive-hermeneutic approach we describe in this book can become a valuable tool as society struggles to cope with the effects of rapid technological change.

Finally, we return to our discussion of what distinguishes smartphones from other modern devices—the fact that we always have them with us. People now spend about a third of their waking hours using smartphones, and that percentage is growing every year.[2] As we outlined throughout the book, these devices are now used to mediate nearly every aspect of our lives. In a very real sense, they are becoming an intrinsic part of our identities. We have argued that digital metaphors, and by extension digital technologies, are changing the meaning of fundamental human concepts. Perhaps the ultimate example of this is the concept of who we are. As we increasingly allow technology to shape who we are, it is imperative that we critically examine the consequences so that we are as well informed about them as we can possibly be.

APPENDIX A

A BRIEF HISTORY OF MEANING

In his influential article "General Semantics," David Lewis argues that meanings have two fundamental aspects.[1] One is context-independent and primarily focused on linguistic structures, and the other takes into account the context in which language is used. Lewis warns of possible problems that might arise by confounding these two aspects and treating them as a single phenomenon. On one hand, the question of meaning can be framed within an abstract and static semantic system—something like understanding the meanings of a language by solely reading one version of a dictionary or analyzing sentences on a blackboard. On the other hand, meaning can be analyzed from the psychological and sociological realities related to the concrete use of an instance of this abstract semantic system. Aware of the necessity to tread lightly to avoid conceptual confusion, we explore both aspects in the following sections of this appendix. Nevertheless, we do not think it is possible to keep these two dimensions completely separate as they intertwine in a dialectical manner. We hope to clarify and leverage this intersection as the argument progresses and use it as one of the building blocks of our ensuing hermeneutic reflections on digital metaphors.

WHAT DOES IT MEAN? OR, THE NONCONTEXTUAL DIMENSION OF MEANING

The non-contextual side of meaning, as we said, concerns analyzing the propositional content of sentences and predications by abstracting them from the context in which they are uttered. This is appealing from the perspective of modern science, since it constrains language to an "objective" analysis separate from the context of use and the psychological and sociological dimensions of its use. In other words, it would be great if we could study sentences in the same way that we study a chemical compound. In this section, we gather some conceptual elements developed over the last century in linguistics and the philosophy of language. These are important pieces of the mosaic we are composing to serve as a conceptual background to our analysis of meaning and metaphors. This process is selective and intentional; we choose only the concepts that are needed to better understand digital metaphors. This necessarily means leaving aside an immense bibliography and the associated long discussions concerning the conceptual nuances surrounding the selected concepts.

The first non-contextual characteristic of meaning that we highlight was originally proposed by Ferdinand de Saussure. Saussure, widely recognized as one of the most prominent linguists of the twentieth century, inspired structuralism, a movement widely supported by the scientific community in the last century.[2] In his *Course in General Linguistics* (1916), Saussure proposed thinking about linguistic systems as consisting of two distinct parts: *langue* and *parole*. In this schema, *langue* is a closed system of signs—each written word in a natural language binds together a concept and an acoustic image.[3] *Parole*, on the other hand, refers to how speakers use a *langue* to communicate messages. The *langue* signs get their meaning from differentiations from other signs within the lexical system. So, someone understands the meaning of "capital" by differentiating it from other signs of the linguistic system like "city" and "town." Such analysis avoids many of the possible problems associated with the temporal nature of language.

Another movement, with very different premises and methods, but that also focuses primarily on the non-contextual aspects of language, is associated with thinkers of the analytical tradition of the philosophy of language, such as Gotlobb Frege, G. E. Moore, Bertrand Russell, and Willard Van Orman Quine. For many, Frege, along with Russell, is one of the precursors of analytic philosophy in the history of twentieth-century thought.[4] His influential work "On Sense and Reference" is a nuanced analysis of some of the non-contextual dimensions of the question of meaning.

Frege's starting point centered around the meaning of a simple linguistic sign. For consistency, let's return to our discussion of Rio. In particular, Frege showed the problems that ensue if all the sign's meanings were contained in its referent. If this were true, then one sign that said "Rio" and another that said "the home of Christ the Redeemer" would mean exactly the same thing, since the statue of Christ the Redeemer is indeed located in Rio. However, it is clear that "the home of Christ the Redeemer" adds something more to the referent than the city name by itself. To solve problems of this nature, Frege postulated that the meaning of words is not limited to the referent of a sign (what is said), but also encompasses sense (how to say what is said), an integral part of the meaning.

When it comes to larger grammatical units, Frege proposed that we must consider sentences based on a new criterion, one linked to the constituent words: "If we now replace one word of the sentence by another having the same referent, but a different sense, this can have no influence upon the referent of the sentence."[5] So, "Rio" and "the home of Christ the Redeemer" must have the same referent, even though the thought or the sense of the sentences is diverse.

Continuing his argumentation in "On Sense and Reference," Frege engages in a bit of poetic analysis by differentiating the aesthetic pleasure linked to the thoughts (or senses) of sentences and the scientific knowledge that would be linked to the sentence referent. He completes this analysis with the following observation: "It is the striving for truth that drives us always to advance from the sense to the referent."[6] Frege then concludes that "we are therefore driven into accepting the truth value of a sentence as its referent. By the truth value of a sentence, I understand the circumstance that it is true or false."[7] That is, the universe of referent possibilities for each and every sentence is the set {True, False}.

This is a paradigmatic statement for analytical philosophy. It concerns the relationships between truth values that, on one hand, abstract the details of each sentence and, on the other, are effectively linked to scientific truths. So a sentence like "Copacabana Beach is located in Rio," could be expressed as a formal function Located_in_Rio (Copacabana) in predicate logic. Meanwhile, "Rio is a city in Brazil," or City_in_Brazil (Rio), could then participate in an equation that would take into account only the referents of both sentences. From a formal perspective, both sentences refer to the same universe of possibilities, true or false. With such a system in place we would be able to make logical inferences using predicate logic, such as the fact that Copacabana Beach is in Brazil.

This definition has significant consequences for the sentences' semantic content, since, as Frege said, "if now the truth value of a sentence is its referent, then on the one hand all true sentences have the same referent and so, on the other hand, do all false sentences. Hence we see that in the referent of the sentence all that is specific is obliterate."[8] Frege's statement illuminates the priority that the analytical tradition within the philosophy of language has given to the logical problems associated with the question of meaning. The standardization of the referent as true or false fit like a glove for modern scholars who sought a precise and unambiguous linguistic approach to describe scientific experiments. To a large extent, this observation about the obliteration of the specific marks a large part of the effort of this philosophical tradition concerning the search for meaning.

Like Frege, Bertrand Russell also thought extensively about the logical aspects of natural languages' meaning as he turned his attention to some aporia of the ordinary use of language. To have a clearer perspective on Russell's approach, let's consider how he would decompose the sentence "the father of bossa nova was from Rio" into separate logical predications.[9] From Russell's point of view, the original sentence would encompass three different and complementary assertions as follows:

1. At least one person is the father of bossa nova.
2. At most one person is the father of bossa nova.
3. Whoever is the father of bossa nova was from Rio.

The original sentence's meaning results from the logical expression that combines each predication using the logical operator "and." Notably, this requires each of the three individual assertions to be true for the whole sentence to be true. This simple example only affords the briefest glimpse into what is a long and rich history of analytical philosophy dealing with this issue.

Our brief encounter with Saussure, Frege, and Russell has revealed some important aspects of the-meaning-of-meaning that are worth highlighting, with a particular focus on what is said. First, according to Saussure's semiotic approach, each semantic unit, such as an individual word, is linked with a multitude of other semantic units in a conceptual lexical network for a given language. Thus, the meaning of a particular word is always related to other words within this linguistic system. To find the meaning of a word is to understand, through a process of comparison, the different predications that are the case for a particular semantic unit, but not for others, within that unit's semantic vicinity within the linguistic system. So, for example, "capital" and "city" are within the same linguistic system, and to understand the meaning of each of these words is to grasp how they differ in their relationships with other units within the conceptual network of the lexicon.

"Capital" is likely to be much more closely linked with words associated with government than "city." Therefore, the meaning of something involves an interplay between approximations and distinctions with other signs in the linguistic system. Such an underlying network of semantic associations will be significant to the understanding of how metaphors are shaped.

While we appreciate the insight of a system of associated signs within a semantic space, we think that only when we get to the level of sentences do the meanings of signs come to fruition and become essential parts of how we see the world. As we discuss in chapter five, the meaning of each sign is actually a collection of associations that are primarily created through the predications attributed to the sign, and such associations are conveyed in language through sentences.

The second point we want to take from this analysis is that meaning is not limited just to the reference of a predication, but also to how it is expressed (sense). Although "Rio" and "the host of the 2016 Summer Olympics" may technically refer to the same place, our choice of which phrase to use brings additional meaning and nuance to the sentence. In both cases the reference (about what) creates a predication involving a physical place in the world, but the sense (what is said) adds another layer of predications internal to the linguistic system depending on the particular choice of words. Using "Rio" emphasizes the city itself, whereas "the host of the 2016 Summer Olympics" shifts the emphasis to the city's role in international sports. Thus, the two expressions do not literally have the same meaning even though they are referring to the same thing. This is because "Rio" is a rich concept whose extent is simply too large to be considered each time it is used. The choice of words is a kind of guide to narrow the extent down to a more manageable size. As we have seen, metaphors are framed in contexts that narrow down the predications we focus on and play an essential part in framing the statement and its meaning.

Third, for some philosophers and linguists, the meaning of

sentences in natural language can be examined first by decomposing their parts into logical predicates, and then by combining the resulting logical expressions in order to determine if the sentence is true or false, i.e., if it refers to real phenomena or not. This type of logical analysis is particularly attractive from the perspective of modern science, since it seeks to be an accurate way of communicating objective descriptions of reality. This logical approach primarily aims to get rid of the ambiguities inherent in natural language. The resulting semantic system would then be purified, rid of the marks of subjectivity and ambiguity. From there it could be used to reference the world in such a way as to contribute precise knowledge of objective realities. The analysis of what is said, when put in terms of expressions whose referents are logical values, true or false, could create the conditions for modern science to express meaning through assertions that could be tested empirically. At the limit, meaning would be reduced to an unequivocal set of logical descriptive assertions whose references to reality could be validated and verified.

Despite its limitations, we want to take two valuable lessons from this logical-analytical approach. First, a natural language sentence potentially contains multiple propositions (X is Y), and each proposition establishes a relation between individuation (X) and predication (is Y). Thus, the meaning of X will be the set of predications (Ys) taken as being the case for a particular speaker or group of speakers of the language. Second, the logical-analytical approach emphasizes empirical verification as the basic criterion for a proposition to have meaning—to the point of excluding propositions that are not verifiable from the universe of meaning. As we shall see later, this limitation of the-meaning-of-meaning does not seem to account for human experiences around meaning, but it points to a relevant dimension of the question about meaning.

At the end of this section, we must ask ourselves if, in addition to these aspects that favor the logical, objective, verifiable, and structural content of meaning, there are no other fundamental characteristics of meaning that only emerge when we analyze the problem in the context of the use of the semantic system, in the way we say what is said through language.

WHAT DO THEY MEAN? OR, THE CONTEXTUAL DIMENSION OF MEANING

So far, we have focused on approaches to meaning that center an effort to logically validate and empirically test meaning. Take the sentence: "The water molecule has two hydrogen atoms and one oxygen atom." This sentence contains the two fundamental aspects of meaning: individuation—a certain substance, in this case, water—and predication that explains its chemical composition. Thus, it is fair to say that part of the meaning we attribute to the linguistic sign of water is this particular chemical composition.

However, these chemical traits are not the only meanings of water and are probably not the most common ways in which we think of water on a daily basis. Indeed, many people are probably not even aware of what the chemical composition of water is. Sentences like "This [liquid] is water," "Water is wonderfully refreshing," and "There is nothing better than water after working out" are more representative of normal conversation. But understanding these sentences and their meanings is challenging. It requires answers to a series of clarifying questions that vary according to the context in which the sentence is uttered: who said it, why they said it, when they said it, to whom, with what intention, etc.

In this section, we investigate another angle of meaning that is related to context. Essential aspects of what we mean by meaning emerge in the concrete ways in which language is used in everyday life. The messiness of reality adds a dense layer of fluid and subjective predications and is almost always ambiguous and imprecise. This ambiguity will then inevitably return us to the arduous task of incessant interpretation of meaning.

The one-time debate between Russell, whose views we just

examined, and P. F. Strawson, two key figures in the philosophy of language, can help us understand this complementary dimension of meaning. William Lycan synthesized their views as follows:

> While Russell thought in terms of sentences taken in the abstract as objects in themselves, and their logical properties in particular, Strawson emphasized how the sentences are used and reacted to by human beings in concrete conversational situations.[10]

And it is with Strawson that we begin to focus on the dimensions of meaning that are directly linked to the context of what is said. In his influential article "On Referring," Strawson emphasizes this dimension:

> The context of utterance is of an importance which it is almost impossible to exaggerate; and by "context " I mean, at least, the time, the place, the situation, the identity of the speaker, the subjects which form the immediate focus of interest, and the personal histories of both the speaker and those he is addresses.[11]

Whereas Russell focused mainly on analyzing meaning from the form of the predications abstracted from their use, Strawson emphasized the context in which sentences are uttered.

Strawson made yet another important contribution to our discussion of metaphors. He related meaning to "the set of rules, habits, conventions for its use in referring."[12] In doing so, he invites us to go beyond the static and self-contained analysis of meaning that we saw with the logical-analytical approach. In doing so we can engage with the fluidity of rules, habits, and linguistic conventions that are always open to revisions and intrinsically linked to the use of language.

But Strawson is certainly not the first great philosopher of language of the last century to highlight the contextual aspect of meaning. In 1954, the journal *Mind* published Strawson's review of Ludwig Wittgenstein's book *Philosophical Investigations*.[13] In the review Strawson noted that Wittgenstein stressed that to understand the meaning of a concept or a word, it is necessary to put it in the linguistic and social context in which it is uttered. Wittgenstein showed the diversity of meanings by examining the simple exclamation "Water!" Depending on the context, this exact sequence of linguistic signs can have any of the meanings: an order, a request, an exclamation, or an answer.

Wittgenstein pointed out that most of the time understanding the "meaning" of a word or concept is linked to its use in language.[14] He suggested that meanings are born and understood within language-games and these, in turn, are expressions that occur within forms of life.[15] So the meaning of words and sentences depends on the type of social context in which speakers create standards and consensus on the meaning of things. Different language games, such as acting out a play," "reporting an event," "giving an order," and "telling a joke," are played in different social contexts. Much of what we call meaning depends on the language-games in which sentences and words are said.

For us, Wittgenstein's approach to meaning points towards at least three important elements. First, a large portion of what we mean by meanings is associated with the language-games in which meaningful words and sentences are spoken. Meaning is associated with the ways we live and interact with each other in the real world. Therefore, meaning is not something that can be dissociated from life—language and life are intrinsically related. However, while Wittgenstein emphasizes the movement from life forms to language-games; we will expand this focus to take into account the dynamic mutuality and bidirectionality between language-games and forms of life over time. Just as life forms shape language-games, language-games affect forms of life. So our discussion of meaning moves beyond the purely epistemological plane (what we understand) and stretches to the deeper waters of the existential and ontological planes (how we live and what we are).

The second source of inspiration that we draw from Wittgenstein has to do with the relationality between entities of meaning. Just as in a game, each piece gets meanings within the relations with other pieces. For instance, in chess, part of the meaning of the knight is that it can pass over any other piece.[16] Meanings are also interrelated in a game of mutual influences, as shown in the cognitive analysis of metaphors. The meanings are not isolated; they can only be captured and understood entirely within semantic networks.

And, finally, the third source of inspiration from *Philosophical Investigations* has to do with the fluidity of the meanings within language-games. Because they are social and temporal, language games are fluid. They change over time, and with them meanings are also refigured and created. As we will see, metaphors are at the heart of this semantic dynamism.

Following this direction of Wittgenstein towards the ordinary use of language as a privileged way to study meanings, J. L. Austin proposed in his theory of speech acts a systematization of the various layers of meaning that emerge from the context of language use.[17] For Austin, if one wants to understand the meaning of a sentence, in addition to understanding its propositional content (what is said), it is necessary to understand its "illocutionary force"—for example, if the sentence is a judgment, a report, an advising, a question, or a warning. The illocutionary force therefore describes the speech act in question, or in other words, what is done when uttering this or that sentence. Austin also adds a third layer of meaning that relates to the effect of what is said on the recipient of the message: the perlocutionary dimension. Hence, the perlocutionary force of a sentence is linked to the effect of persuasion, fear, and repudiation with the harmony that a sentence generates. Thus, in order to understand the meaning of a sentence, it is also necessary to consider its illocutionary (what is done when saying such a thing) and perlocutionary (what the sentence causes in the audience) force.

Another essential aspect, one foreshadowed in the title of Austin's key work *How to do Things with Words* and explored more extensively by John Searle, is that even when we are declaring something, e.g. "this is water," we are changing our relationship with the world, because we can now act differently towards that substance and possibly drink it. So, it's not just typical performative sentences like "I promise to repay this loan" that modify the state of things, but also those called constatives—propositions declaring something to be the case. This observation is especially relevant in the context of metaphors because it highlights that the predications suggested by metaphors have pragmatic effects—their meanings involve a different way of seeing and acting in the world.

Speech act theory emphasizes that meanings are not limited to the simple description of things; meanings affect reality. To understand the meaning of something implies in many cases to act in a certain way. As we have already seen with digital metaphors, the demonstration that a metaphor's meaning has been understood is an effective interaction with the digital artifact that implements the metaphor. The meaning of a message like "Accept Friendship Invitation?" is understood through an interaction associated with a change in the state of affairs. This pragmatic dimension of language, which opened up a very extensive field of study, is fundamental to our research since it highlights the intrinsic relationship between meaning and acting in the world.

At the end of this detour by a philosophical-linguistic analysis of meaning, we need to take stock of the mosaic of tiles that we have gathered, since they form the backdrop of our discussion on how metaphors—particularly digital ones—create and modify meanings. One of our goals is to draw the reader's attention to the complexity of defining meaning. Something that we take for granted, when put under an analytical light, is revealed to be full of nuances and difficulties in understanding. Thus, we metaphorically speak of a mosaic to emphasize the notion of an approximation, where some pieces still need conceptual polish before there is a single continuous image. But we'll see that such a mosaic works very well as a background from which we can talk more deeply about metaphors. Keeping that in mind, we discuss five tiles of this mosaic that are particularly relevant to us.

First, meaning is a set of predications directed towards a concept, real or abstract. Thus, the meaning of "Rio" involves predications as "a beautiful city," "a city of contrast," and "the place where bossa nova was born." So to understand how new digital technologies affect meanings, we need to explore how they change and create predications for concepts like communication, friendship, dating, public debate, buying, selling, and so on.

Second, meanings are not just linguistic entities that are only of interest to philosophers and cognitive scientists. Meanings refer to things and phenomena in the world and, as we discuss throughout Part I, they mediate how we relate to others, the world and ourselves. The predications that dominate the meanings associated with Rio have direct consequences on the propensity for tourists to visit the city, for example. To understand the meaning is to look at something based on certain predications. Thus, I can relate to the city of Rio de Janeiro as "a beautiful city" or as "a city of contrasts," or in both ways. The number of predications packed into a concept is such that there are typically too many predications to have them all at once. Meanings are, therefore, fundamental to how we live and act; meanings matter. And it is precisely because digital technologies affect meanings that digital technologies matter.

Third, as we saw through the different angles of the linguistic system proposed by the structuralists, and then by the language-games suggested by Wittgenstein, meanings do not occur in isolation. To understand the meaning of something is to understand a network of meaning and associations within contexts that guide interpretation. The creation of metaphors depends on this interconnected web of meanings and how the understanding of metaphors will vary depending on the contexts in which they are employed.

Fourth, from Frege's discussions of sense and reference, we learn that meaning is not just "what" we refer to with a particular linguistic expression, but that "how" we refer also affects meaning. To say that something is an "optical surface positioning device" or to say it is a "mouse" has different meanings, even if both expressions refer to the same physical entity.

And fifth, meanings are not static; they are closely linked to the use that speakers make of language, the ways we live, and social and cultural structures. As we saw in Tara Westover's story, meanings can vary radically from one person to another or even within one person's life, depending on the particular experiences of the speakers. Westover's autobiography (and possibly all biographies) is a journey through new meanings mediated by her educational trajectory. Meanings are malleable, and any person's dominant predications of a concept will evolve over time, overlapping and overriding previous predications in a constant game of sedimentation and innovation. This book explores the role of digital technologies in creating new forms of life, affecting predications and triggering semantic innovations that alter the meaning of concrete and abstract concepts of everyday life.

APPENDIX B

KEY TERMS

attention economy: The idea that a significant portion of the economy, including many apps featured in this book, is now devoted to capturing and holding our attention.

cognitive association: When concepts are experienced close together in time, they tend to become linked in our minds. If it happens enough, thinking about one of the concepts will naturally lead to thinking about the other concept. Thus, when we hear someone say "A B C D E F," our mind naturally expects them to say "G" next.

cognitive-hermeneutic approach: An epistemological approach that draws from cognitive science and hermeneutics' conceptual frameworks to analyze digital technologies.

cognitive-hermeneutic cycle: A process in which semantic horizons are expanded through an interpretative process of new meanings.

digital metaphor: A metaphor used by the developers of a technology to explain what it is to users. A critical difference from regular metaphors is that digital metaphors are given materiality by being implemented.

digital technology: A technology that requires or embeds a digital processor. Computational hardware and software are typical examples of digital technologies.

directed attention: Attention that requires effort. This type of attention is typically deployed in service of our intentions.

directed attention fatigue: Since directed attention requires effort, it involves deploying resources in a way that can become fatigued. Such fatigue has a negative impact not only on the ability to deploy attention, but also on cognition and health.

Hebb's rule: A neural learning rule first proposed by Donald Hebb. The rule states that, when the firing of one neuron facilitates the firing of another neuron, the connection between them is strengthened such that when the first neuron fires in the future, it will be more efficient in helping the second neuron to fire. This is often phrased as "fire together, wire together" in neuroscience.

involuntary attention: Attention that requires no effort. This type of attention is associated with things that are so compelling that they are difficult to ignore.

impertinent predication: A predication that is not part of the lexicon of a natural language. This is a basic mechanism used by metaphors to suggest new meanings.

life-world: A complex and multifaceted structure that constitutes how we understand our experiences in the world through the integration of various concepts and predications that constitute our semantic networks.

metaphorical spiral: A model that describes how our horizons of meaning are changed by metaphors over time.

predication: The process of associating attributes, characteristics, and relations to one individual or category. It is the predicative part of a proposition.

prefigured horizon of meaning: A horizon of meaning before encountering a semantic innovation, such as a metaphor.

refigured horizon of meaning: A horizon of meaning altered by a semantic innovation, such as a metaphor.

semantic horizon or horizon of meaning: The set of cultural and
  fundamental knowledge that a person or group of people
  has before encountering a new technology; their "semantic
  universe," roughly a combination of their experience and their
  knowledge

semantic innovation: The creation of new meanings. We
  suggested that metaphors are a paradigm of semantic
  innovations.

technology: "A set of practices humans use to transform the
  material world, practices involved in creating and using
  material things."[1]

Notes

INTRODUCTION

1. Elizabeth Palermo, "Who Invented the Lightbulb?," Live Science, November 23, 2021, https://www.livescience.com/43424-who-invented-the-light-bulb.html.

2. Margaret Mead and Rhoda Metraux, "Image of the Scientist among High-School Students," *Science* 126, no. 3270 (1957): 384–90.

3. Sarah Harmon and Katie McDonough, "The Draw-A-Computational-Creativity-Researcher Test (DACCRT): Exploring Stereotypic Images and Descriptions of Computational Creativity," in *Proceedings of the 10th International Conference on Computational Creativity*, ed. Kazjon Grace, Michael Cook, Dan Ventura, and Mary Lou Maher, 243–49 (Charlotte, NC: Association for Computational Creativity, 2019).

4. Eric Schatzberg, *Technology: Critical History of a Concept* (Chicago: University of Chicago Press, 2018), 2.

5. Luciano Floridi, *The Fourth Revolution: How the Infosphere Is Reshaping Human Reality* (Oxford: Oxford University Press, 2014), 25–27. As Schatzberg notes in his critical history of the concept, "the definition of technology is a mess." As described by Paul Forman and other philosophers of science and technology, its relationship to similar concepts such as science, applied science, and engineering is complex and motivated by diverse social forces. Jennifer Daryl Slack and J. Macgregor Wise also highlight that most people today relate the concept of technology (or technologies in the plural) to artifacts created by the technological process. See Paul Forman, "The Primacy of Science in Modernity, of Technology in Postmodernity, and of Ideology in the History of Technology," *History and Technology* 23, no. 1–2 (March 2007): 1–152; Jennifer Daryl Slack and J. Macgregor Wise, *Culture and Technology: A*

Primer, 2nd ed. (New York: Peter Lang, 2005), 95–99. This book proposes a reflection that makes clear the complexity of the technological process that is hidden behind and within the various technologies (technological artifacts).

6. What Floridi refers to as second- and third-order technologies. *Fourth Revolution*, 25–27. We will not consider the implications of third-order technologies, such as internet of things devices, but we will assume that even if humans are not directly connected to this technology, they are "removed from the loop," as Floridi suggests; they are still aware that their refrigerator automatically ordered milk and a drone delivered it to their front door.

7. James Burke, *Connections: Alternative History of Technology* (New York: Macmillan, 1980).

8. Christian Katzenbach and Stefan Larsson, "How Metaphors Shape the Digital Society," *Digital Society Blog*, Alexander von Humboldt Institute for Internet and Society, 2018, https://www.hiig.de/en/dossier/how-metaphors-shape -the-digital-society/.

9. Don Norman, *The Design of Everyday Things*, rev. and exp. ed. (New York, Basic Books, 2013), 3.

10. We propose this concept along the lines of Gadamer's famous hermeneutical notion of "fusion of horizons." Hans-Georg Gadamer, *Truth and Method* (New York: Crossroad, 1989).

11. For those wondering, we will discuss this metaphor in more detail, especially in chapter six.

12. Michael W. Morris et al., "Metaphors and the Market: Consequences and Preconditions of Agent and Object Metaphors in Stock Market Commentary," *Organizational Behavior and Human Decision Processes* 102, no. 2 (March 2007): 174–92.

13. Michael L. Slepian et al., "Shedding Light on Insight: Priming Bright Ideas," *Journal of Experimental Social Psychology* 46, no. 4 (July 2010): 696–700.

14. Stephen Kaplan, "The Expertise Challenge," in *Fostering Reasonableness: Supportive Environments for Bringing Out Our Best*, ed. Rachel Kaplan and Avik Basu (Ann Arbor, MI: Maize Books, 2015), 50.

15. Erin Woo, "QR Codes Are Here to Stay. So Is the Tracking They Allow," *New York Times*, July 26, 2021, https://www.nytimes.com/2021/07/26/technology /qr-codes-tracking.html.

16. The COVID-19 pandemic has increased the visibility and use of QR codes, but they require an immense effort of communication and explanation for the general public to engage with them. See for instance Victoria Turk, "In a Touch-Free World, the QR Code Is Having Its Moment," Wired, August 18, 2020, https://www.wired.com/story/in-a-touch-free-world-the-qr-code -is-having-its-moment/.

17. Albert Borgmann, *Technology and the Character of Contemporary Life: A Philosophical Inquiry* (Chicago: University of Chicago Press, 1987).

18. Donald A. Schön, "Generative Metaphor: A Perspective on Problem-Setting in Social Policy," in *Metaphor and Thought*, 2nd ed., ed. Andrew Ortony (Cambridge: Cambridge University Press, 1979), 141.

19. Marcus Jahnke, "Revisiting Design as a Hermeneutic Practice: An Investigation of Paul Ricoeur's Critical Hermeneutics," *Design Issues* 28, no. 2 (2012): 39.

20. Orson Scott Card, *Ender's Game* (New York: Tor Books, 2014), 249.

21. George Lakoff and Mark Johnson, *Metaphors We Live By* (Chicago: University of Chicago Press, 1980).

CHAPTER ONE

1. MacMillan Dictionary, "Common Metaphors," *MacMillan Dictionary* (blog), May 11, 2011, https://www.macmillandictionaryblog.com/common-metaphors.

2. For an extensive treatment of this and many other aspects of metaphors' importance, see James Geary, *I Is an Other: The Secret Life of Metaphor and How It Shapes the Way We See the World* New York: HarperCollins, 2011).

3. "Metaphor in one form or another is absolutely fundamental to the way language systems develop over time and are structured, as well as to the way human beings consolidate and extend their ideas about themselves, their relationships and their knowledge of the world." Low et al. eds., *Researching and Applying Metaphor in the Real World* (Amsterdam: John Benjamins, 2010), xii.

4. For instance, Don Ihde, *Technology and the Lifeworld: From Garden to Earth* (Bloomington: University of Indiana Press, 1990). See Carl Mitcham, *Thinking Through Technology: The Path Between Engineering and Philosophy* (Chicago: University of Chicago Press, 1994).

5. Mark Coeckelbergh, "Language and Technology: Maps, Bridges, and Pathways," *AI & Society* 32, no. 2 (May 2017): 186.

6. Aristotle, *Poetics*, trans. Stephen Halliwell, Loeb Classical Library 23 (Cambridge, MA: Harvard University Press, 1995), 115.

7. "Thus, poetry and oratory mark out two distinct universes of discourse. Metaphor, however, has a foot in each domain. With respect to structure, it can really consist in just one unique operation, the transfer of the meanings of words; but with respect to function, it follows the divergent destinies of oratory and tragedy. Metaphor will therefore have a unique structure but two

functions: a rhetorical function and a poetic function." Paul Ricoeur, *The Rule of Metaphor: Multi-Disciplinary Studies of the Creation of Meaning in Language* (Toronto: University of Toronto Press, 1978), 12.

8. "We all naturally find it agreeable to get hold of new ideas easily: words express ideas, and therefore those words are the most agreeable that enable us to get hold of new ideas. Now strange words simply puzzle us; ordinary words convey only what we know already; it is from metaphor that we can best get hold of something fresh. When the poet calls old age "a withered stalk," he conveys a new idea, a new fact, to us by means of the general notion (genous) of 'lost bloom.'" Aristotle, *Art of Rhetoric*, trans. J. H. Freese, Loeb Classical Library 193 (Cambridge, MA: Harvard University Press, 1926), 397.

9. Aristotle, *Poetics*, 115.

10. John T. Kirby notes that Aristotle recognizes an important use of metaphors of naming things that do not have proper names of their own. This is certainly a perfect match for naming things that are designed and created through digital techniques. "Aristotle on Metaphor," *American Journal of Philology* 118, no. 4 (1997): 517–54.

11. For a theory of how this might work, see Eric Chown, "Spatial Prototypes," in *Spatial Behavior and Linguistic Representation*, ed. Thora Thenbrink, Jan Wiener, and Christophe Claramunt, 97–114 (Oxford: Oxford University Press, 2013).

12. Ivor Armstrong Richards, *The Philosophy of Rhetoric* (Oxford: Oxford University Press, 1936); Ivor Armstrong Richards, *Principles of Literary Criticism* (London: Routledge, 2017); Max Black, "More About Metaphor," *Dialectica* 31, no. 34 (1977): 431–57; Max Black, *Models and Metaphors: Studies in Language and Philosophy* (Ithaca, NY: Cornell University Press, 1962); Monroe C. Beardsley, "The Metaphorical Twist," *Philosophy and Phenomenological Research* 22, no. 3 (1962): 293–307; Monroe C. Beardsley, "Metaphorical Senses," *Noûs* 12, no. 1 (1978): 3–16; Paul Ricoeur, "The Metaphorical Process as Cognition, Imagination, and Feeling," *Critical Inquiry* 5, no. 1 (October 1978): 143–59; Paul Ricoeur, "Metaphor and the Central Problem of Hermeneutics," *Graduate Faculty Philosophy Journal* 3, no. 1 (April 1973): 42–58; Ricoeur, *Rule of Metaphor: Multi-Disciplinary Studies.*

13. Richards, *Philosophy of Rhetoric.*

14. Black, *Models and Metaphors.*

15. Paul Ricoeur, *The Rule of Metaphor: The Creation of Meaning in Language* (London: Psychology Press, 2003).

16. Judith Oster, "Frost's Poetry of Metaphor," in *The Cambridge Companion to Robert Frost*, ed. Robert Fagen (Cambridge: Cambridge University Press, 2001), 155.

17. Lakoff and Johnson, *Metaphors We Live By*.

18. Markus Tendahl and Raymond W. Gibbs Jr., "Complementary Perspectives on Metaphor: Cognitive Linguistics and Relevance Theory," *Journal of Pragmatics* 40, no. 11 (November 2008): 1823–64.

19. For example, Eleanor Rosch, "Principles of Categorization," *Concepts: Core Readings*, ed. Eric Margolis and Stephen Laurence, 251–70 (Cambridge, MA: MIT Press, 1999).

20. See, for example, Raymond W. Gibbs Jr., *The Cambridge Handbook of Metaphor and Thought* (Cambridge: Cambridge University Press, 2008).

21. Keith J. Holyoak and Dušan Stamenković, "Metaphor Comprehension: A Critical Review of Theories and Evidence," *Psychological Bulletin* 144, no. 6 (June 2018): 641–71; Walter Kintsch, "Metaphor Comprehension: A Computational Theory," *Psychonomic Bulletin & Review* 7, no. 2 (June 2000): 257–66.

22. Paola de Varona, "5 Experts Explain mRNA Vaccines for Non-Science People," Verywell Health, December 21, 2020, https://www.verywellhealth.com/expla ining-mrna-vaccines-experts-social-media-5092888.

23. George Lakoff, "The Neural Theory of Metaphor," in *The Cambridge Handbook of Metaphor and Thought*, ed. Raymond W. Gibbs Jr., 17–38 (New York: Cambridge University Press, 2008).

24. Lynne Cameron, "Metaphor and Talk," in *The Cambridge Handbook of Metaphor and Thought*, ed. Raymond W. Gibbs Jr., 197–211 (New York: Cambridge University Press, 2008).

25. Jonathan H. Blavin and I. Glenn Cohen, "Gore, Gibson, and Goldsmith: The Evolution of Internet Metaphors in Law and Commentary," *Harvard Journal of Law and Technology* 16, no. 1 (2002): 266.

26. Blavin and Cohen, "Gore, Gibson, and Goldsmith," 267.

27. "Most importantly, there is also significant research indicating the prominence of metaphor in many areas of abstract thought and in people's emotional and aesthetic experiences." Gibbs Jr., *Cambridge Handbook of Metaphor and Thought*, 3.

28. Lakoff, "Neural Theory of Metaphor."

29. "Metaphors carry not only ideational content but also something of speakers' attitudes and values in respect of that content. Vehicle choice offers affective potential." Cameron, "Metaphor and Talk," 203.

30. Cameron, "Metaphor and Talk."

31. Adam Alter, *Irresistible: The Rise of Addictive Technology and the Business of Keeping Us Hooked* (New York: Penguin, 2018).

32. "In August 2012, Netflix introduced a subtle new feature called post-play. With post-play, a 13-episode season of *Breaking Bad* became a single 13-hour film. As one episode ended, the Netflix player automatically loaded the next

one, which began playing 5 seconds if the previous episode left you with a cliffhanger, all you had to do was sit still as the next episode began and the cliffhanger resolved itself. Before August 2012, you had to decide the next episode; now, you have to decide not to watch the next episode." Alter, *Irresistible*, 208.

33. "The FrameWorks Institute tested a series of metaphors and found that the metaphors 'rampant versus regular CO_2,' 'osteoporosis of the sea,' and 'climate's heart' were most successful at building an understanding of climate change." Anne K. Armstrong, "Using Metaphor and Analogy in Climate Change Communication," in *Communicating Climate Change: A Guide for Editors*, ed. Anne K. Armstrong, Marianne E. Krasny, and Jonathan P. Schuldt (Ithaca, NY: Cornell University Press, 2019), 70.

34. Mark Coeckelbergh and Wessel Reijers, "Narrative Technologies: A Philosophical Investigation of the Narrative Capacities of Technologies by Using Ricoeur's Narrative Theory," *Human Studies* 39, no. 3 (September 2016): 325–46.

35. Coeckelbergh and Reijers, "Narrative Technologies," 336–37.

36. See, for instance, Yossi Sheffi's article describing the concerns of Uber drivers about the possible consequences of a negative review. "A Failure to Treat Workers with Respect Could Be Uber's Achilles' Heel," MIT Technology Review, September 22, 2014, https://www.technologyreview.com/2014/09/22/12491/a-failure-to-treat-workers-with-respect-could-be-ubers-achilles-heel/.

37. On the role of metaphors for framing public policies see Schön, "Generative Metaphor."

38. George M. Marakas, Richard D. Johnson, and Jonathan W. Palmer, "A Theoretical Model of Differential Social Attributions toward Computing Technology: When the Metaphor Becomes the Model," *International Journal of Human-Computer Studies* 52, no. 4 (April 2000): 719–50.

39. In this book we end each chapter with a short section that serves to put key ideas and quotes in one place in a kind of summary. In this, we have taken inspiration from the recent books of Nobel Prize-winner Daniel Kahneman, such as *Thinking, Fast and Slow* (New York: Farrar, Strauss and Giroux, 2011), 35; and Daniel Kahneman, Olivier Sibony, and Cass R. Sunstein, *Noise: A Flaw in Human Judgment* (New York: Little, Brown, 2021).

CHAPTER TWO

1. Sherry Turkle, *Alone Together: Why We Expect More from Technology and Less from Each Other* (London: Hachette UK, 2017).
2. Richard Ling, *Taken for Grantedness: The Embedding of Mobile Communication into Society* (Cambridge, MA: MIT Press, 2012).
3. Margaret E. Morris, *Left to Our Own Devices: Outsmarting Smart Technology to Reclaim Our Relationships, Health, and Focus* (Cambridge, MA: MIT Press, 2018).
4. Elaine Reese and Robyn Fivush, "The Development of Collective Remembering," *Memory* 16, no. 3 (April 2008): 201–12.

CHAPTER THREE

1. William F. Buckley Jr., "Do We Really Need Home Computers?," *On the Right* (syndicated column), June 19, 1982.
2. Norman, *Design of Everyday Things*.
3. It is important to note that in this text we use the term appropriation in the technical sense of the German hermeneutical tradition of the original term *Aneignung*. *Aneignen* means to make one's own what was initially alien and is applied here to the process in which a cultural artifact (text, monument, technological artifact) gains meaning for anyone who engages with the artifact, and the meaning proposed by the artifact becomes part of how they relate to the world. There is therefore no connotation with an unjust or illegal usurpation of something that belonged to the other and is forcibly taken away.
4. Paul Ricoeur, *Hermeneutics and the Human Sciences: Essays on Language, Action and Interpretation* (Cambridge, Cambridge University Press, 1981), 185.
5. Austin Carr, "Will Apple's Tacky Design Philosophy Cause A Revolt?," Fast Company, September 11, 2012, https://www.fastcompany.com/1670760/will-apples-tacky-software-design-philosophy-cause-a-revolt.
6. David Oswald, "Affordances and Metaphors Revisited: Testing Flat vs. Skeuomorph Design with Digital Natives and Digital Immigrants," in *Proceedings of the 32nd International BCS Human Computer Interaction Conference (HCI)*, 1–11. Swindon, UK: BCS Learning & Development, 2018; Huifeng Jin, "Influence of Icon Design Style on User's Cognition," in *Proceedings of the 6th International Conference on Humanities and Social Science Research (ICHSSR 2020)*, ed. Xuemei Du, Chunyan Huang, and Yulin Zhong, 550–53 (Paris: Atlantis Press, 2020).

7. Marianne V. T. van den Boomen, *Transcoding the Digital: How Metaphors Matter in New Media* (Amsterdam: Institute of Network Cultures, 2014).

8. van den Boomen, *Transcoding the Digital*, 12.

9. Paul M. Leonardi, "Digital Materiality? How Artifacts without Matter, Matter," *First Monday* 15, no. 6 (June 2010), https://doi.org/10.5210/fm.v15i6.3036.

10. Although it seems more appropriate to think specifically about the logical structures of the software, because the way these structures are physically implemented obviously involves some sort of matter (bits are ultimately pulses of electrical energy).

11. Bruno Latour, "On Technical Mediation," *Common Knowledge* 3, no. 2 (1994): 29–64.

12. Latour, "On Technical Mediation," 38.

13. van den Boomen, *Transcoding the Digital*.

14. van den Boomen, *Transcoding the Digital*; N. Katherine Hayles, *Writing Machines* (Cambridge, MA: MIT Press, 2002); Lakoff and Johnson, *Metaphors We Live By*.

15. van den Boomen, *Transcoding the Digital*, 188.

16. van den boomen, *Transcoding the Digital*, 188.

17. Hayles, *Writing Machines*.

18. Ricoeur, *Rule of Metaphor: Multi-Disciplinary Studies*, 115.

CHAPTER FOUR

1. We recognize that some filmmakers like to tinker with their films even after release, hence the explosion of "director's cuts" and other special editions.

2. The metaphorical spiral that we suggest is inspired by Ricoeur's triple mimesis, which we discuss in chapter five. Paul Ricoeur, *Time and Narrative*, vol. 1 (Chicago, University of Chicago Press, 1984), 52*ff*.

3. Facebook, "A/B Tests Types Available on Facebook," Meta, accessed July 30, 2021, https://www.facebook.com/business/.

4. Ray C. He, "Introducing New Like and Share Buttons," *Facebook for Developers* (blog), November 6, 2013, https://developers.facebook.com/blog/post/2013/11/06/introducing-new-like-and-share-buttons/.

5. ExplainingComputers, "5 Most Annoying Computing Things!," June 20, 2021, video, 13:15, https://www.youtube.com/watch?v=HuF4NfUok3Q.

6. Herbert A. Simon, "Designing Organizations for an Information-Rich World," In *Computers, Communications, and the Public Interest*, ed. M. Greenberger, 37–52 (Baltimore, MD: John Hopkins University Press, 1971).

7. Michael H. Goldhaber, "The Attention Economy and the Net," *First Monday* 2, no. 4 (April 7, 1997), https://doi.org/10.5210/fm.v2i4.519.

8. BBC News, "US Officer Plays Taylor Swift Song to Try to Block Video," BBC, July 2, 2021, https://www.bbc.com/news/technology-57698858; Lorch, "How Children Are Spoofing Covid-19 Tests with Soft Drinks," BBC, July 6, 2021, https://www.bbc.com/future/article/20210705-how-children-are-spoofing-covid-19-tests-with-soft-drinks; PairPlayApp, "Pairplay," accessed August 17, 2021, https://www.pairplayapp.com.

9. Edward Tenner, *Why Things Bite Back: Technology and the Revenge of Unintended Consequences* (New York: Vintage Books, 1997); Edward Tenner, *Our Own Devices: How Technology Remakes Humanity* (New York: Vintage Books, 2004); Morris, *Left to Our Own Devices*.

10. Annette Markham, "The Limits of the Imaginary: Challenges to Intervening in Future Speculations of Memory, Data, and Algorithms," *New Media & Society* 23, no. 2 (February 2021): 382–405.

11. S. O'Dea, "Mobile Device Daily Usage Time in U.S. 2014-2021," Statista, February 27, 2020, https://www.statista.com/statistics/1045353/mobile-device-daily-usage-time-in-the-us/.

12. Alter, *Irresistible*.

13. Simon Kemp, "Digital 2021: Global Overview Report," Datareportal, January 27, 2021, https://datareportal.com/reports/digital-2021-global-overview-report.

14. Joanna C. Yau and Stephanie M. Reich, "Buddies, Friends, and Followers: The Evolution of Online Friendships," in *Online Peer Engagement in Adolescence: Positive and Negative Aspects of Online Interaction*, ed. Nejra Van Zalk and Claire P. Monks, 18–34 (London: Routledge, 2020), https://doi.org/10.4324/9780429468360-2.

15. dana boyd, *It's Complicated: The Social Lives of Networked Teens* (New Haven, CT: Yale University Press, 2014), 94–95.

16. Maryam Mohsin, "10 Tiktok Statistics that You Need to Know in 2020," Oberlo, February 16, 2021, https://www.oberlo.com/blog/tiktok-statistics.

17. Jane Wakefield, "People Devote Third of Waking Time to Mobile Apps," BBC, January 12, 2022, https://www.bbc.com/news/technology-59952557.

18. Jeff Desjardins, "How Long Does It Take to Hit 50 Million Users," Visual Capitalist, 2018, https://www.visualcapitalist.com/how-long-does-it-take-to-hit-50-million-users/.

19. Skye Gould, "It Took 75 Years for the Telephone to Reach 100 Million Users and It Took Candy Crush 15 Months," Business Insider, July 28, 2015, https://www.businessinsider.com/it-took-75-years-for-the-telephone-to-reach-100-million-users-and-it-took-candy-crush-15-months.

20. Patrick Collison (@patrickc) , "Fun fact: We deployed 3,350 new versions of the Stripe API last year," Twitter, March 7, 2021, https://twitter.com/patrickc /status/1368692809436303360.

21. Ricoeur, "Metaphorical Process as Cognition, Imagination, and Feeling."

22. In this way "mouse" is actually an unusual tech metaphor that reflects its origins from pure research labs at the Stanford Research Institute. The name "mouse" tells us only about the device's size and shape (and that once upon a time the cord looked like a tail), not about its usage. We are not asked to believe that the pointing device is a rodent.

CHAPTER FIVE

1. Augustine, *Confessions. Books 9–13*, trans. W. Watts, Loeb Classic Library 26 (Cambridge, MA: Harvard University Press, 1912), vol. 26, bk. 11.

2. Tara Westover, *Educated: A Memoir* (New York: Random House, 2018).

3. In Appendix A we offer a brief account of the history of meaning as a concept. In doing so we choose a few critical stepping stones, drawn from linguistics and hermeneutics, that might offer interested readers a deeper understanding of some of the more applied notions developed in this book.

4. Paul Ricoeur, *Interpretation Theory: Discourse and the Surplus of Meaning* (Fort Worth: Texas Christian University Press, 1976), 11.

5. "Whereas the genuinely logical subject is the bearer of a singular identification, what the predicate says about the subject can always be treated as a 'universal' feature of the subject. Subject and predicate do not do the same job in the proposition. The subject picks out something singular—Peter, London, this table, the fall of Rome, the first man who climbed Mt. Everest, etc.—by means of several grammatical devices which serve this logical function: proper names, pronouns, demonstratives (this and that, now and then, here and there, tenses of the verb as related to the present), and 'definite descriptions' (the so and so). What they all have in common is that they all identify one and only one item. The predicate, in contrast, designates a kind of quality, a class of things, a type of relation, or a type of action." Ricoeur, *Interpretation Theory*, 9.

6. Ernst Cassirer, *An Essay on Man* (New Haven, CT: Yale University Press, 2021).

7. "Reason is a very inadequate term with which to comprehend the forms of man's cultural life in all their richness and variety. But all these forms are symbolic forms. Hence, instead of defining man as an animal rationale, we should define him as an animal symbolicum. By so doing we can designate

his specific difference, and we can understand the new way open to man—the way to civilization." Cassirer, *Essay on Man*, 26.

8. Consider the curious practice in which the President of the United States pardons a Thanksgiving turkey. Thus, a turkey's life is spared, a turkey that would otherwise be eaten as part of a unique celebration of the nation's origins. It is a curious example of how certain types of food become intertwined with cultural traditions in multiple layers over time. Betty C. Monkman, "Pardoning the Thanksgiving Turkey," White House Historical Association, accessed August 3, 2021, https://www.whitehousehistory.org/pardoning-the-thanksgiving-turkey.

9. See, for instance, Thomas A. Vilgis, "Evolution—Culinary Culture—Cooking Technology," in *Culinary Turn: Aesthetic Practice of Cookery*, ed. Nicolaj van der Meulen and Jörg Wiesel, 149–60 (Bielefeld: Transcript Verlag, 2017).

10. Nathaniel Popper, "As Diners Flock to Delivery Apps, Restaurants Fear for Their Future," *New York Times*, June 9, 2020, https://www.nytimes.com/2020/06/09/technology/delivery-apps-restaurants-fees-virus.html.

11. Bryan Miller, "How Mobile Technology Is Changing the Way We Dine Out," WSJ Online, October 25, 2013. https://www.wsj.com/articles/SB10001424052702303680404579143903633457212.

12. P. Brennan, "Sexual Selection," *Nature Education Knowledge* 3, no. 10 (2010): 79.

13. Ernst Cassirer, "A Clue to the Nature of Man: The Symbol," in *An Essay on Man: An Introduction to a Philosophy of Human Culture* (New Haven, CT: Yale University Press, 1992), 41–44.

14. "United States: Online Dating Users in the U.S. 2017-2024," Statista, July 5, 2021, https://www.statista.com/statistics/417654/us-online-dating-user-numbers/.

15. Miller et al., "Online and Offline Relationships," in *How the World Changed Social Media*, 100–113 (London: UCL Press, 2016).

16. Giulia Ranzini and Christoph Lutz, "Love at First Swipe? Explaining Tinder Self-presentation and Motives," *Mobile Media & Communication* 5, no. 1 (2017): 81; Sherry Turkle, *Reclaiming Conversation: The Power of Talk in a Digital Age* (New York: Penguin, 2016), 130.

17. Matthew H. Rafalow and Britni L. Adams, "Digitally Mediated Connections and Relationship Persistence in Bar Settings," *Symbolic Interaction* 40, no. 1 (2017): 25–42.

18. Edwin J. C. van Leeuwen, Katherine A. Cronin, and Daniel B. M. Haun, "Tool Use for Corpse Cleaning in Chimpanzees," *Scientific Reports* 7 (March 2017): article 44091; Jessica Pierce, "What the Grieving Mother Orca Tells Us about

How Animals Experience Death," The Conversation, August 24, 2018, http:// theconversation.com/what-the-grieving-mother-orca-tells-us-about-how -animals-experience-death-101230.

19. David Rowell, "Dead Musicians Are Taking the Stage Again in Hologram Form. Is This the Kind of Encore We Really Want?," *Washington Post*, October 30, 2019, https://www.washingtonpost.com/magazine/2019/10/30/dead-mus icians-are-taking-stage-again-hologram-form-is-this-kind-encore-we-really -want/.

20. Dustin L. Abramson and Joseph Johnson Jr., Creating a Conversational Chat Bot of a Specific Person, US Patent 10,853,717, filed April 11, 2017, and issued December 1, 2020, https://patentimages.storage.googleapis.com/8d/2a/7e/32 5266284d79df/US10853717.pdf.

21. Johann Michel, *Homo Interpretans: Towards a Transformation of Hermeneutics*, trans. David Pellauer (Lanham, MD: Rowman & Littlefield, 2019).

22. We use the term "life-world" mainly related to Husserl's phenomenology (*Lebenswelt* in German) in the sense of a reservoir of meaning as suggested by Paul Ricoeur. "The return to the *Lebenswelt* can more effectively play this paradigmatic role for hermeneutics if the *Lebenswelt* is not confused with some sort of ineffable immediacy and is not identified with the vital and emotional envelope of human experience, but rather is construed as desig- nating the reservoir of meaning, the surplus of sense in living experience, which renders the objectifying and explanatory attitude possible." *From Text to Action: Essays in Hermeneutics II*, trans. John B. Thompson and Kathleen Blamey (New York: Bloomsbury, 2008), 2:86.

23. Black, *Models and Metaphors*, 242.

24. Black, *Models and Metaphors*, 243. See also Francesca Ervas, Elisabetta Gola, and Maria Grazia Rossi, eds., *Metaphor in Communication, Science and Edu- cation* (Berlin: de Gruyter, 2017).

25. Sally Wyatt, "Danger! Metaphors at Work in Economics, Geophysiology, and the Internet," *Science, Technology & Human Values* 29, no. 2 (April 2004): 244.

26. Schön, "Generative Metaphor," 140.

27. Fernanda Abreu, "Rio 40 Graus," *SLA 2 Be Sample*, EMI Records Brasil, 1992.

28. Ricoeur explores the creative and imaginative aspects of metaphors through the dialectics of sense (what is said) and reference (that about which some- thing is said). The reference is the projection of language toward the world of meanings. For Ricoeur, metaphors open up a new form of metaphorical reference, as they do not point towards something that is already there, but they point to some new possible meanings. They suggest an expansion of one's current set of meanings, as they "redescribe reality inaccessible to direct description" *Time and Narrative*, vol. 1, xi.

29. Aristotle, *Poetics*, 115.

30. Predication in the grammatical sense of attributing characteristics to a term.

31. Ricoeur, "Metaphorical Process as Cognition, Imagination, and Feeling.".

32. In the terminology of Jean Cohen, "Structure du Langage Poétique," *Les Etudes Philosophiques* 22, no. 2 (1967): 219.

33. Black, "More About Metaphor."

34. "ARGUMENT IS WAR metaphor that expressions from the vocabulary of war, e.g., 'attack a position', 'indefensible', 'strategy', 'new line of attack', 'win', 'gain ground', etc. form a systematic way of talking about the battling aspects of arguing." Lakoff and Johnson, *Metaphors We Live By*, 456.

35. Ricoeur, *Time and Narrative*, vol. 1, xi; Ricoeur, "Metaphor and the Central Problem of Hermeneutics." Ricoeur approximates metaphor's creative power with the idea of mimesis. In turn, this provides an interesting perspective on metaphors through the analysis of the moments of the narrative triple mimesis: prefiguration, configuration, and refiguration. The implicative complexes that are part of a metaphorical statement exist in a prefigured world of meaning characterized by their established literal meanings. A new metaphor is a twist of the prefigured meaning in a configured impertinent predication that references an emergent meaning. Readers of the metaphor refigure their world through the tensional fusion of horizons between the metaphorical reference suggested by the metaphor and their prefigured literal meanings involved in the metaphor. We could call this process the metaphorical cycle.

36. Ricoeur, *Time and Narrative*, vol. 1.

CHAPTER SIX

1. Jeff Hawkins, *A Thousand Brains: A New Theory of Intelligence* (New York: Basic Books, 2021).

2. Dylan Meconis, Scott McCloud, and Syne Mitchell, "Learning Machine Learning," Google, 2019, https://cloud.google.com/products/ai/ml-comic-1

3. Randall Munroe, *Thing Explainer: Complicated Stuff in Simple Words* (London: Hachette UK, 2015).

4. Scott McCloud, "Google Chrome," Google, 2008, https://www.google.com/googlebooks/chrome/.

5. Personal communication, September 2020.

6. Olaf Sporns, Giulio Tononi, and Rolf Kötter, "The Human Connectome: A Structural Description of the Human Brain," *PLoS Computational Biology* 1, no. 4 (2005): e42, https://doi.org/10.1371/journal.pcbi.0010042.

7. Donald O. Hebb, *The Organization of Behavior* (New York: Wiley, 1949).
8. William James, *Psychology: The Briefer Course*, ed. Gordon Allport (New York: Harper & Row, 1961).
9. James, *Psychology: The Briefer Course*, 256.
10. Kahneman, *Thinking, Fast and Slow*.
11. Robert D. Oades, "The Role of Noradrenaline in Tuning and Dopamine in Switching between Signals in the CNS," *Neuroscience and Biobehavioral Reviews* 9, no. 2 (June 1985): 261–82.
12. The relationship between long-term memory and arousal has been known for more than fifty years. What is interesting is that the relationship can actually be negative in the short term. See, for example, Lewis J. Kleinsmith and Stephen Kaplan, "Paired-Associate Learning as a Function of Arousal and Interpolated Interval," *Journal of Experimental Psychology* 65, no. 2 (February 1963): 190–93; Lewis J. Kleinsmith and Stephen Kaplan, "Interaction of Arousal and Recall Interval in Nonsense Syllable Paired-Associate Learning," *Journal of Experimental Psychology* 67, no. 2 (February 1964): 124–26.). This is so counterintuitive that numerous replicative studies were done, all of which confirmed the original findings. Almost forty years later, a model was published that supported Kleinsmith and Kaplan's original hypothesis that neural fatigue was responsible for the effect. Eric Chown, "Reminiscence and Arousal: A Connectionist Model," *Proceedings of the Annual Meeting of the Cognitive Science Society* 24 (2002), https://escholarship.org/content/qt4v45r3gd/qt4v45r3gd.pdf.
13. Antonio R. Damasio, *Descartes' Error: Emotion, Reason and the Human Brain* (New York: Random House, 2006), 49.
14. Sergio Sismondo, *Science without Myth: On Constructions, Reality, and Social Knowledge* (Albany: State University of New York Press, 1996).
15. Mark J. Landau, Brian P. Meier, and Lucas A. Keefer, "A Metaphor-Enriched Social Cognition," *Psychological Bulletin* 136, no. 6 (November 2010): 1045–67.
16. Lawrence E. Williams and John A. Bargh, "Keeping One's Distance: The Influence of Spatial Distance Cues on Affect and Evaluation," *Psychological Science* 19, no. 3 (March 2008): 302–8.
17. Lawrence E. Williams and John A. Bargh, "Experiencing Physical Warmth Promotes Interpersonal Warmth," *Science* 322, no. 5901 (October 24, 2008): 606–7.
18. Eric J. Ivancich, "Fostering Clarity to Achieve Reasonableness," in *Fostering Reasonableness: Supportive Environments for Bringing Out Our Best*, ed. Rachel Kaplan and Avik Basu (Ann Arbor, MI: Maize Publishing, 2015), https://doi.org/10.3998/maize.13545970.0001.001.

19. Stephen Kaplan, "Beyond Rationality: Clarity-Based Decision Making," in *Environment, Cognition, and Action: An Integrated Approach*, ed. Thomas Garling and Gary W. Evans, 171–90 (Oxford: Oxford University Press, 1991); Stephen Kaplan, "Attention and Fascination: The Search for Cognitive Clarity," in *Humanscape: Environments for People*, ed. Stephen Kaplan and Rachel Kaplan, 84–93 (Belmont, CA: Duxbury, 1978).

20. Ivancich, "Fostering Clarity to Achieve Reasonableness."

21. This is often referenced as 7+-2, but in 1975 George Mandler reanalyzed the data and determined that the actual "magic number" is 5+-2. George Mandler, "Consciousness: Respectable, Useful, and Probably Necessary," in *Information Processing and Cognition*, ed. Robert L. Solso (Hillsdale, NJ: Lawrence Erlbaum, 1975).

22. James, *Psychology: The Briefer Course.*

23. Kahneman, *Thinking, Fast and Slow*, 35.

24. For a review, see Stephen Kaplan and Marc G. Berman, "Directed Attention as a Common Resource for Executive Functioning and Self-Regulation," *Perspectives on Psychological Science* 5, no. 1 (January 2010): 43–57.

25. Roy F. Baumeister, Mark Muraven, and Dianne M. Tice, "Ego Depletion: A Resource Model of Volition, Self-Regulation, and Controlled Processing," *Social Cognition* 18, no. 2 (June 2000): 130–50.

26. Bernadine Cimprich, "Development of an Intervention to Restore Attention in Cancer Patients," *Cancer Nursing* 16, no. 2 (April 1993): 83–92.

27. Goldhaber, "Attention Economy and the Net."

28. Avik Basu, Jason Duvall, and Rachel Kaplan, "Attention Restoration Theory: Exploring the Role of Soft Fascination and Mental Bandwidth," *Environment and Behavior* 51, no. 9–10 (November 2019): 1055–81.; Kaplan and Berman, "Directed Attention as a Common Resource"; Rachel Kaplan, "Attention! That's a Precious Resource," *Our Planet* 2017, no. 2 (May 2018): 28–29, https://doi.org/10.18356/63bb4c1f-en.

29. See, for example, Marcel Adam Just et al., "A Neurosemantic Theory of Concrete Noun Representation Based on the Underlying Brain Codes," *PLOS One* 5, no. 1 (January 13, 2010): article e8622.

30. Kintsch, "Metaphor Comprehension."

31. *Troilus and Cressida*, ed. Barbar A. Mowat and Paul Werstine (Washington, DC: Folger Shakespeare Library, 2007), 3.3.150–155.

32. See, for example, Christian Lachaud, "Conceptual Metaphors and Embodied Cognition: EEG Coherence Reveals Brain Activity Differences between Primary and Complex Conceptual Metaphors during Comprehension," *Cognitive Systems Research* 22–23 (June 2013): 12–26; Valentina Bambini et

al., "Decomposing Metaphor Processing at the Cognitive and Neural Level through Functional Magnetic Resonance Imaging," *Brain Research Bulletin* 86, no. 3-4 (October 10, 2011): 203–16; Zohar Eviatar and Marcel Adam Just, "Brain Correlates of Discourse Processing: An fMRI Investigation of Irony and Conventional Metaphor Comprehension," *Neuropsychologia* 44, no. 12 (June 23, 2006): 2348–59.

33. For example, Isabel C. Bohrn, Ulrike Altmann, and Arthur M. Jacobs, "Looking at the Brains behind Figurative Language—a Quantitative Meta-Analysis of Neuroimaging Studies on Metaphor, Idiom, and Irony Processing," *Neuropsychologia* 50, no. 11 (September 2012): 2669–83.

34. Kenneth J. Craik, "The Nature of Explanation," *Journal of Philosophy* 40, no. 24 (1943): 661, https://doi.org/10.2307/2018933.61.

35. It must also be noted that we are ignoring a large class of metaphors that are not directly relevant to this book, notably spatial metaphors such as "success is up," "the future is in front," and many others. Such metaphors are discussed in great detail in Lakoff and Johnson, *Metaphors We Live By*; Chown, "Spatial Prototypes"; Mark Johnson, *The Body in the Mind: The Bodily Basis of Meaning, Imagination, and Reason* (Chicago: University of Chicago Press, 1987).

36. Gerard J. Steen et al., "Metaphor in Usage," *Cognitive Linguistics* 21, no. 4 (November 2010): 765–96.

37. Morris et al., "Metaphors and the Market."

38. Landau, Meier, and Keefer, "A Metaphor-Enriched Social Cognition."

39. Sindya Bhanoo, "Who Should Translate Amanda Gorman's Work? That Question Is Ricocheting around the Translation Industry," *Washington Post,* March 24, 2021, https://www.washingtonpost.com/entertainment/books/book-translations-gorman-controversy/2021/03/24/8ea3223e-8cd5-11eb-9423-04079921c915_story.html; Alex Marshall, "Amanda Gorman's Poetry United Critics. It's Dividing Translators," *New York Times*, March 26, 2021, https://www.nytimes.com/2021/03/26/books/amanda-gorman-hill-we-climb-translation.html.

40. Amanda Gorman (@TheAmandaGorman), Twitter Timeline, https://twitter.com/TheAmandaGorman.

CHAPTER SEVEN

1. Don Ihde, *Philosophy of Technology: An Introduction* (St. Paul, MN: Paragon House, 1998).

2. Artificial intelligence has been a vital enabler of changing the boundaries

between human and technological agency and accountability. Mark Coeckelbergh, *AI Ethics* (Cambridge, MA: MIT Press, 2020).

3. See for instance Chrysoula Gatsou, "The Importance of Metaphors for User Interaction with Mobile Devices," in *Design, User Experience, and Usability: Users and Interactions*, ed. Aaron Marcus, 520–29 (Cham: Springer International Publishing, 2015); Javier Carbonell, Antonio Sánchez-Esguevillas, and Belén Carro, "The Role of Metaphors in the Development of Technologies. The Case of the Artificial Intelligence," *Futures* 84 (November 2016): 145–53; Thomas D. Erickson, "Working with Interface Metaphors," in *Readings in Human–Computer Interaction: Toward the Year 2000*, 2nd ed., ed. Ronald M. Baecker et. al, 147–51 (San Francisco: Morgan Kaufmann, 1995); Thomas P. Moran and Shumin Zhui, "Beyond the Desktop Metaphor in Seven Dimensions," in *Designing Integrated Digital Work Environments: Beyond the Desktop Metaphor*, ed. Victor Kaptelinin and Mary Czerwinski, 335–54 (Cambridge, MA: MIT Press, 2007).

4. Paul Ricoeur, "The Power of Speech: Science and Poetry," *Philosophy Today* 29, no. 1 (February 1985): 59–70.

5. Coeckelbergh, and Reijers, "Narrative Technologies."

6. Coeckelbergh and Reijers, "Narrative Technologies"

7. Michel, *Homo Interpretans*.

8. The next three chapters are greatly expanded versions of work that first appeared in Eric Chown and Fernando Nascimento, "Software and Metaphors: The Hermeneutical Dimensions of Software Development," in *Interpreting Technology*, ed. Wessel Reijers, Alberto Romele, and Mark Coeckelbergh, 229–48 (Lanham, MD: Rowman & Littlefield, 2021).

CHAPTER EIGHT

1. Horace Dediu (@asymco), "Touch UIs are now enabling $1 Trillion of economic activity. it's time to see what's next. Apple is about to launch a new iPhone," Twitter, October 13, 2020, https://twitter.com/asymco/status/13160 6009453107201.

2. Mic Wright, "The Original iPhone Announcement Annotated," The Next Web, September 9, 2015, https://thenextweb.com/apple/2015/09/09/genius -annotated-with-genius/.

3. "Virtual Scroll Wheel Patent Shows Alternate iOS Input Method," AppleInsider, August 7, 2012, https://appleinsider.com/articles/12/08/07/virtual_scr oll_wheel_patent_shows_alternate_ios_input_method.

4. Wright, "The Original iPhone Announcement Annotated."
5. Lakoff and Johnson, *Metaphors We Live By*.
6. For a rundown, see Seth Fiegerman, "The Experts Speak: Here's What People Predicted Would Happen When the iPhone Came Out," Business Insider, June 29, 2012, https://www.businessinsider.com/iphone-predictions-from-20 07-2012-6.
7. P. Daniel, "Four Years of Disruption: Cell Phone Industry Financials 2007-2011," PhoneArena, July 14, 2011, https://www.phonearena.com/news/Four -years-of-disruption-cell-phone-industry-financials-2007-2011_id20153.
8. Phil B., "Importance of First – Why Nokia Cares about Super Affordable Mobile Phones," "Importance of First – Why Nokia Cares about Super Affordable Mobile Phones," Nokia Conversations: The Official Nokia Blog, September 16, 2013, https://web.archive.org/web/20130916190754/http://conversatio ns.nokia.com/2013/09/13/importance-of-first-why-nokia-cares-about-super -affordable-mobile-phones.
9. Jason Aten, "14 Years Ago, Steve Jobs Sent the Most Important Email in the History of Business," Inc., June 5, 2021, https://www.inc.com/jason-aten/14 -years-ago-steve-jobs-sent-most-important-email-in-history-of-business .html.

CHAPTER NINE

1. The exception to this is games where "in-app purchase" economies thrive. Many games are built such that users can pay to make them easier to play, or to upgrade skills, or even to avoid seeing ads.
2. Goldhaber, "Attention Economy and the Net."
3. Thomas H. Hutchinson, *Here Is Television, Your Window to the World* (New York: Hastings House, 1950).
4. Benedict Evans (@benedictevans), "The average FB user is eligible to see 1500-2000 items a day. This sounds absurd, but it's only 150 friends posting, liking, sharing 10 times a day each. And 3 seconds x 1500 items is 75 minutes. There will always be a filter - the only question is what kind," Twitter, January 23, 2018, https://twitter.com/benedictevans/status/955625564945096709.
5. dana boyd, *It's Complicated: The Social Lives of Networked Teens* (New Haven, CT: Yale University Press, 2014), 47–49.
6. Benedict Evans, "Death of the Newsfeed," *Benedict Evans* (blog), April 3, 2018, https://www.ben-evans.com/benedictevans/2018/4/2/the-death-of-the-new sfeed.

7. Alter, *Irresistible*, 3.

8. Neil Postman, *Amusing Ourselves to Death: Public Discourse in the Age of Show Business* (New York: Penguin, 1985), 5

9. He, "Introducing New Like and Share Buttons."

CHAPTER TEN

1. Mark Milian, "Why Text Messages Are Limited to 160 Characters," *Los Angeles Times Blog*, May 3, 2009, https://latimesblogs.latimes.com/technology/2009/05/invented-text-messaging.html.

2. It is intriguing that the underlying idea of spatial distortion is reflected in the Greek etymology of these three technologies: tele-vision, tele-gramma, tele-phone. So, these technologies ask users to see them as "seeing," "writing," and "talking," but from afar. Together they form a "cluster" of metaphors around tele-communication.

3. Chris Gayomali, "The Text Message Turns 20: A Brief History of SMS," The Weekly Law Reports, 2012, https://theweek.com/articles/469869/text-mess age-turns-20-brief-history-sms; Christine Erickson, "A Brief History of Text Messaging," Mashable, September 21, 2012, http://mashable. com/2012/09/21/text-Messaging-History.

4. Turkle, *Alone Together.*

CHAPTER ELEVEN

1. Reese and Fivush, "Development of Collective Remembering."

2. "Watch & Manage Your Memories," Google Photos Help, Google, accessed August 6, 2021, https://support.google.com/photos/answer/9454489?co=GE NIE.Platform%3DAndroid&hl=en.

3. Lauren Goode, "I Called Off My Wedding. The Internet Will Never Forget," Wired, April 6, 2021, https://www.wired.com/story/weddings-social-media -apps-photos-memories-miscarriage-problem/.

4. Melissa Fay Greene, "You Won't Remember the Pandemic the Way You Think You Will," *The Atlantic*, April 6, 2021, https://www.theatlantic.com/magazine /archive/2021/05/how-will-we-remember-covid-19-pandemic/618397/.

5. Ulric Neisser and Nicole Harsch, "Phantom Flashbulbs: False Recollections of Hearing the News about Challenger," *Affect and Accuracy in Recall: Studies of "Flashbulb" Memories*, ed. Eugene Winograd and Ulric Neisser, 9–31 (Cambridge: Cambridge University Press, 1992).

1. For example, Adrian Kingsley-Hughes, "Who Do I Pay to Get the 'Phone' Removed from My iPhone?" ZDNet, April 12, 2021, https://www.zdnet.com /article/who-do-i-pay-to-get-the-phone-removed-from-my-iphone/.

2. Kevin Roose, "Why Are We Still Calling the Things in Our Pockets 'Cell Phones'?" Intelligencer, June 24, 2014, https://nymag.com/intelligencer/2014 /06/why-are-we-still-calling-them-cell-phones.html.

3. Irfan Ahmad, "60+ Fascinating Smartphone Apps Usage Statistics for 2019," Social Media Today, March 23, 2019, https://www.socialmediatoday.com/ne ws/60-fascinating-smartphone-apps-usage-statistics-for-2019-infographic /550990/.

4. Stephanie Chan, "Global Consumer Spending in Mobile Apps Reached a Record $111 Billion in 2020, up 30% from 2019," Sensor Tower, January 4, 2021, https://sensortower.com/blog/app-revenue-and-downloads-2020.

5. Joanna Stern, "iPhone? AirPods? MacBook? You Live in Apple's World. Here's What You Are Missing," *Wall Street Journal*, June 4, 2021, https://www.wsj .com/articles/iphone-airpods-macbook-you-live-in-apples-world-heres-wh at-you-are-missing-11622817653.

6. John Gruber, "Theme Parks and Public Parks," Daring Fireball, June 6, 2021, https://daringfireball.net/linked/2021/06/06/theme-parks-and-public-parks.

7. Dieter Bohn, "Apple Isn't Just a Walled Garden, It's a Carrier," The Verge, June 7, 2021, https://www.theverge.com/2021/6/7/22521476/apple-walled-garden -carrier-app-store-innovation.

8. Cory Doctorow, "The Coming Civil War over General Purpose Computing," Boing Boing, August 23, 2012, https://boingboing.net/2012/08/23/civilwar .html.

9. Doctorow, "Coming Civil War."

10. Wakefield, "People Devote Third of Waking Time to Mobile Apps."

11. Alex Heath, "Facebook Plans First Smartwatch for Next Summer with Two Cameras, Heart Rate Monitor," The Verge, June 9, 2021, https://www.theve rge.com/2021/6/9/22526266/facebook-smartwatch-two-cameras-heart-rate -monitor.

12. Amber Neely, "TikTok Can Collect Users' Biometric Data According to Its New Privacy Policy," AppleInsider, June 4, 2021, https://appleinsider.com/art icles/21/06/04/tiktok-can-collect-users-biometric-data-according-to-its-new -privacy-policy.

13. Chase Buckle, "Which Smartphone Features Really Matter to Consumers?" GWI, January 9, 2019, https://blog.gwi.com/chart-of-the-week/smartphone -features-consumers/.

CHAPTER THIRTEEN

1. Alma Whitten and J. D. Tygar, "Why Johnny Can't Encrypt: A Usability Evaluation of PGP 5.0," in *Security and Usability: Designing Secure Systems That People Can Use*, ed. Lorrie Faith Cranor and Simson Garfinkel, 679–702 (Sebastopol, CA: O'Reilly, 2005).

2. Albese Demjaha et al., "Metaphors Considered Harmful? An Exploratory Study of the Effectiveness of Functional Metaphors for End-to-End Encryption," in *Proceedings of 2018 Workshop on Usable Security (USEC)* (Reston, VA: Internet Society, 2018).

3. Whitten and Tygar, "Why Johnny Can't Encrypt," 684.

4. Leontine Jenner, "Backdoor: How a Metaphor Turns into a Weapon," *Digital Society Blog*, Alexander von Humboldt Institute for Internet and Society, 2018, https://www.hiig.de/en/backdoor-how-a-metaphor-turns-into-a-weapon/.

5. Rong Yin and Haosheng Ye, "The Black and White Metaphor Representation of Moral Concepts and Its Influence on Moral Cognition," *Acta Psychologica Sinica* 46, no. 9 (September 2014): 1331.

6. "Project Implicit," Harvard University, accessed August 5, 2021, https://implicit.harvard.edu/implicit/index.jsp.

7. For a discussion, see Keith Payne, Laura Niemi, and John M. Doris, "How to Think about 'Implicit Bias,'" *Scientific American*, March 27, 2018, https://www.scientificamerican.com/article/how-to-think-about-implicit-bias/.

8. Landau, Meier, and Keefer, "A Metaphor-Enriched Social Cognition."

CHAPTER FOURTEEN

1. They account for 15 percent of the S&P and 30 percent of the Nasdaq. See Adam Levy, "Investing in FAANG or MAMAA Stocks," The Motley Fool, updated July 21, 2022, https://www.fool.com/investing/stock-market/market-sectors/information-technology/faang-stocks/.

2. Conor Cawley, "10 Stats That Prove Silicon Valley Hasn't Fixed Its Diversity Problem," Tech.co, June 25, 2018, https://tech.co/news/stats-silicon-valley-fixed-diversity-problem-2018-06.

3. Adam D. I. Kramer, Jamie E. Guillory, and Jeffrey T. Hancock, "Experimental Evidence of Massive-Scale Emotional Contagion through Social Networks," *Proceedings of the National Academy of Sciences of the United States of America* 111, no. 24 (June 17, 2014): 8788–90.

4. Mike Monteiro, *Ruined by Design: How Designers Destroyed the World, and What We Can Do to Fix It* (Independently published, 2019), 125.

5. Isabella Kwai, "Consumer Groups Target Amazon Prime's Cancellation Process," *New York Times*, January 14, 2021. https://www.nytimes.com/2021/01/14/world/europe/amazon-prime-cancellation-complaint.html.

6. Nir Eyal, "The *New York Times* Uses the Very Dark Patterns It Derides," *Nirandfar* (blog), May 30, 2021, https://www.nirandfar.com/cancel-new-york-times/.

CHAPTER FIFTEEN

1. John Gramlich, "10 Facts about Americans and Facebook," Pew Research, June 1, 2021, https://www.pewresearch.org/fact-tank/2021/06/01/facts-about-americans-and-facebook/.

2. Tom Webster, "Subscription Confusion, and the State of Podcasting Data," *I Hear Things* (blog), February 26, 2021, https://tomwebster.substack.com/p/subscription-confusion-and-the-

3. James Cridland, "'Follow Our Podcast': Apple Podcasts to Stop Using 'Subscribe,'" Podnews, March 9, 2021, https://podnews.net/update/follow-not-subscribe.

4. Safiya Umoja Noble, *Algorithms of Oppression: How Search Engines Reinforce Racism* (New York: New York University Press, 2018); Ruha Benjamin, *Race After Technology: Abolitionist Tools for the New Jim Code* (Medford, MA: Polity Press, 2019).

5. Caroline Criado Perez, *Invisible Women: Data Bias in a World Designed for Men* (New York: Abrams, 2019).

6. Grant Blank, "The Digital Divide Among Twitter Users and Its Implications for Social Research," *Social Science Computer Review* 35, no. 6 (December 2017): 679–97.

7. Andrea Rosales and Mireia Fernández-Ardèvol, "Ageism in the Era of Digital Platforms," *Convergence* 26, no. 5–6 (December 2020): 1074–87.

8. Cathy O'Neil, *Weapons of Math Destruction: How Big Data Increases Inequality and Threatens Democracy* (New York: Crown, 2016).

9. See, for example, Khari Johnson, "This New Way to Train AI Could Curb Online Harassment," Wired, August 4, 2021, https://www.wired.com/story/new-way-train-ai-curb-online-harassment/. For a more academic take, see Maarten Sap et al., "The Risk of Racial Bias in Hate Speech Detection," in *Proceedings of the 57th Annual Meeting of the Association for Computational Linguistics*, ed. Preslav Nakov and Alexis Palmer, 1668–78 (Florence, Italy: Association for Computational Linguistics, 2019).

10. Ben Collins and Brandy Zadrozny, "Anti-Vaccine Groups Changing into 'Dance Parties' on Facebook to Avoid Detection," NBC News, July 22, 2021, https://www.nbcnews.com/tech/tech-news/anti-vaccine-groups-changing -dance-parties-facebook-avoid-detection-rcna1480."

11. See, for example, Oscar Schwartz, "In 2016, Microsoft's Racist Chatbot Revealed the Dangers of Online Conversation," IEEE Spectrum. November 25, 2019, https://spectrum.ieee.org/in-2016-microsofts-racist-chatbot-reveal ed-the-dangers-of-online-conversation.

12. Sara Wachter-Boettcher, *Technically Wrong: Sexist Apps, Biased Algorithms, and Other Threats of Toxic Tech* (New York: W. W. Norton, 2017), 10–11.

13. Benjamin, *Race After Technology*.

14. Jeremy N. Bailenson, "Nonverbal Overload: A Theoretical Argument for the Causes of Zoom Fatigue," *Technology, Mind, and Behavior* 2, no. 1 (2021). https://doi.org/10.1037/tmb0000030.

15. Stuart A. Thompson and Charlie Warzel, "Opinion: How Facebook Became a Tool of the Far Right," *New York Times*, January 14, 2021, https://www.nyti mes.com/2021/01/14/opinion/facebook-far-right.html.

16. William J. Brady et al., "How Social Learning Amplifies Moral Outrage Expression in Online Social Networks," *Science Advances* 7, no. 33 (August 2021), https://doi.org/10.1126/sciadv.abe5641.

17. Joanna Stern, "Investigation: How TikTok's Algorithm Figures out Your Deepest Desires," *Wall Street Journal,* July 21, 2021, https://on.wsj.com/3hR 6GzA.

18. For both a review and a theoretical perspective, see Kaplan and Berman, "Directed Attention as a Common Resource."

19. Kaplan and Berman, "Directed Attention as a Common Resource."

20. Basu, Duvall, and Kaplan, "Attention Restoration Theory."

CHAPTER SIXTEEN

1. Wyatt, "Danger! Metaphors at Work"; Sally Wyatt, "Metaphors in Critical Internet and Digital Media Studies," *New Media & Society* 23, no. 2 (February 2021): 406–16.

2. Blavin and Cohen, "Gore, Gibson, and Goldsmith."

3. This double movement is analogous to the complementary aspects of interpretation (hermeneutics) as proposed by Paul Ricoeur. "Hermeneutics appeared henceforth as a battle field traversed by two opposing trends, the first tending toward a reductive explanation, the second tending toward a

recollection or a retrieval of the original meaning of the symbol." *Rule of Metaphor*, 376.

4. Robert Frost, "Education by Poetry," in *The Selected Prose of Robert Frost*, ed. Hyde Cox and Edward Connery Lathem (New York: Holt, Rinehart and Winston, 1966), 35.

5. Oster, "Frost's Poetry of Metaphor."

6. "To conclude, the light of human minds is perspicuous words, but by exact definitions first snuffed, and purged from ambiguity; reason is the pace; increase of science, the way; and the benefit of mankind, the end. And on the contrary, metaphors, and senseless and ambiguous words, are like ignes fatui; and reasoning upon them, is wandering amongst innumerable absurdities; and their end, contention, and sedition, or contempt [indifference]." Thomas Hobbes, *Leviathan*, ed. J. C. A. Gaskin (Oxford: Oxford University Press, 1998), 32.

7. For instance, see Andreas Musolff, *Metaphor and Political Discourse: Analogical Reasoning in Debates about Europe* (Basingstoke: Palgrave Macmillan, 2004); J. Charteris-Black, *Politicians and Rhetoric: The Persuasive Power of Metaphor* (London: Palgrave Macmillan, 2011).

8. Andrew Goatly, *Washing the Brain: Metaphor and Hidden Ideology* (Metaphor and Hidden Ideology. Amsterdam: John Benjamins Publishing, 2007).

9. Katzenbach and Larsson, "How Metaphors Shape the Digital Society."

10. Goatly, *Washing the Brain*.

11. *As You Like It*, ed. Alan Brissenden (Oxford: Oxford University Press, 1993), 2.7.145–149.

12. Tom Simonite, "Facebook's Like Buttons Will Soon Track Your Web Browsing to Target Ads," MIT Technology Review, September 16, 2015, https://www.technologyreview.com/2015/09/16/166222/facebooks-like-buttons-will-soon-track-your-web-browsing-to-target-ads/.

13. Yau and Reich, "Buddies, Friends, and Followers."

14. According to Roberto Simanowski, this redescription of the concept of friendship was one of Facebook's foundational ideas. *Facebook Society: Losing Ourselves in Sharing Ourselves* (New York: Columbia University Press, 2018), 6.

15. "Welcome to Junto," Junto, accessed August 5, 2022, https://learn.junto.foundation/.

16. The English equivalent would be "together," triggering associations of equal participation and collaborative engagement.

17. "Terminology," Junto, accessed August 5, 2022, https://learn.junto.foundation/the-app/terminology.

18. Tim Hwang and Karen Levy, "'The Cloud' and Other Dangerous Metaphors,"

The Atlantic, January 20, 2015, https://www.theatlantic.com/technology/arch
ive/2015/01/the-cloud-and-other-dangerous-metaphors/384518/.

19. See, for instance, Brianna Abbott, "Google AI Beats Doctors at Breast Cancer
Detection—Sometimes," *Wall Street Journal*, January 1, 2020, https://www.wsj
.com/articles/google-ai-beats-doctors-at-breast-cancer-detectionsometimes
-11577901600.

20. See, for instance, Nicola Jones, "How to Stop Data Centres from Gobbling up
the World's Electricity," *Nature* 561, no. 7722 (September 2018): 163–66.

21. Regarding recent discussions on targeted advertising trends see, for instance,
Jiwoong Shin and Jungju Yu, "Targeted Advertising and Consumer Inference," SSRN, September 7, 2020, https://doi.org/10.2139/ssrn.3688258; Natasha Lomas, "Targeted Ads Offer Little Extra Value for Online Publishers,
Study Suggests," TechCrunch, May 31, 2019, http://techcrunch.com/2019/05
/31/targeted-ads-offer-little-extra-value-for-online-publishers-study-sugge
sts/; Araw Mahdawi, "Targeted Ads Are One of the World's Most Destructive
Trends. Here's Why," *The Guardian*, November 5, 2019, http://www.theguar
dian.com/world/2019/nov/05/targeted-ads-fake-news-clickbait-surveillance
-capitalism-data-mining-democracy.

22. Tim O'Reilly, "Data Is the New Sand," The Information, February 24, 2021,
https://www.theinformation.com/articles/data-is-the-new-sand?utm_sourc
e=ti_app.

CONCLUSION

1. Wyatt, "Metaphors in Critical Internet and Digital Media Studies."
2. Wakefield, "People Devote Third of Waking Time to Mobile Apps."

APPENDIX A

1. David Lewis, "General Semantics," in *Montague Grammar*, ed. Barbara H. Partee, 1–50 (New York: Academic Press, 1976).

2. Carol Sanders, "Introduction: Saussure Today," in *The Cambridge Companion
to Saussure*, ed. Carol Sanders, 1–6 (Cambridge: Cambridge University Press,
2004).

3. "Language is a system of signs that expresses ideas." Ferdinand Mongin de
Saussure, Course in General Linguistics (New York: Columbia University
Press, 2011), 16.

4. "Bertrand Russell and Gottlob Frege are the two giants on whose shoulders

analytic philosophy rests." Michael Beaney, "Russell and Frege," in *The Cambridge Companion to Bertrand Russell*, ed. Nicholas Griffin (Cambridge Cambridge University Press, 2003), 128.

5. Gottlob Frege, "Sense and Reference," *The Philosophical Review* 57, no. 3 (1948): 214.
6. Frege, "Sense and Reference," 216.
7. Frege, "Sense and Reference," 216.
8. Frege, "Sense and Reference," 217.
9. Russell's original sentence was "The author of Waverley was Scotch."
10. William G. Lycan, *Philosophy of Language: A Contemporary Introduction*, 3rd ed. (London: Routledge, 2018), 19.
11. P. F. Strawson, "On Referring," *Mind* 59, no. 235 (July 1950): 328.
12. Strawson, "On Referring," 328.
13. P. F. Strawson, "VI—Critical Notice," *Mind* 63, no. 249 (January 1954): 70–99.
14. "For a large class of cases of the employment of the word 'meaning'—though not for all—this word can be explained in this way: the meaning of a word is its use in the language." Ludwig Wittgenstein, *Philosophical Investigations*, 4th ed., ed. P. M. S. Hacker and Joachim Schulte, trans. G. E. M. Anscombe, P. M. S. Hacker, and Joachim Schulte (Malden, MA: Wiley-Blackwell, 2009), 43.
15. "The word 'language-game' is used here to emphasize the fact that the speaking of language is part of an activity, or of a form of life." Wittgenstein, *Philosophical Investigations*, 15c.
16. The knight is the only chess piece that can "jump over" other pieces, irrespective of whether those pieces are black or white. It is also the only piece that moves in an L-shape form, creating a unique set of possible interactions with other game pieces. To understand the full meaning of the knight, one needs to understand how it is used to "play the game" of chess; it is from the game dynamics that the meaning of each piece emerges.
17. John Langshaw Austin, *How to Do Things with Words* (Oxford: Clarendon Press, 1975). Later significantly developed by John R. Searle in *Speech Acts: An Essay in the Philosophy of Language* (Cambridge: Cambridge University Press, 1969).

APPENDIX B

1. Schatzberg, *Technology*, 2.

Bibliography

Abbott, Brianna. "Google AI Beats Doctors at Breast Cancer Detection—Sometimes." *Wall Street Journal.* January 1, 2020. https://www.wsj.com/articles/google-ai-beats-doctors-at-breast-cancer-detectionsometimes-11577901600.

Abramson, Dustin I., and Joseph Johnson Jr. Creating a Conversational Chat Bot of a Specific Person. US Patent 10,853,717, filed April 11, 2017, and issued December 1, 2020. https://patentimages.storage.googleapis.com/8d/2a/7e/32 5266284d79df/US10853717.pdf.

Abreu, Fernanda. "Rio 40 Graus." *SLA 2 Be Sample.* EMI Records Brasil, 1992.

Ahmad, Irfan. "60+ Fascinating Smartphone Apps Usage Statistics for 2019." Social Media Today. March 23, 2019. https://www.socialmediatoday.com/news/60-fascinating-smartphone-apps-usage-statistics-for-2019-infographic/55 0990/.

Alter, Adam. *Irresistible: The Rise of Addictive Technology and the Business of Keeping Us Hooked.* New York: Penguin, 2018.

AppleInsider. "Virtual Scroll Wheel Patent Shows Alternate iOS Input Method." AppleInsider. August 7, 2012. https://appleinsider.com/articles/12/08/07/virtual_scroll_wheel_patent_shows_alternate_ios_input_method.

Aristotle. *Art of Rhetoric.* Translated by J. H. Freese. Loeb Classical Library 193. Cambridge, MA: Harvard University Press, 1926.

———. *Poetics.* Translated by Stephen Halliwell. Loeb Classical Library 23. Cambridge, MA: Harvard University Press, 1995.

Armstrong, Anne K. "Using Metaphor and Analogy in Climate Change Communication." In *Communicating Climate Change: A Guide for Educators*, edited by Anne K. Armstrong, Marianne E. Krasny, and Jonathon P. Schuldt, 70–74. Ithaca, NY: Cornell University Press, 2019.

Aten, Jason. "14 Years Ago, Steve Jobs Sent the Most Important Email in the His-

tory of Business." Inc. June 5, 2021. https://www.inc.com/jason-aten/14-yea
rs-ago-steve-jobs-sent-most-important-email-in-history-of-business.html.

Augustine. *Confessions Books 9–13*. Translated by W. Watts. Loeb Classical Library
26. Cambridge, MA: Harvard University Press, 1912.

Austin, John Langshaw. *How to Do Things with Words*. Oxford: Clarendon Press,
1975.

Bailenson, Jeremy N. "Nonverbal Overload: A Theoretical Argument for the
Causes of Zoom Fatigue." *Technology, Mind, and Behavior* 2, no. 1 (2021).
https://doi.org/10.1037/tmb0000030.

Bambini, Valentina, Claudio Gentili, Emiliano Ricciardi, Pier Marco Bertinetto,
and Pietro Pietrini. "Decomposing Metaphor Processing at the Cognitive
and Neural Level through Functional Magnetic Resonance Imaging." *Brain
Research Bulletin* 86, no. 3–4 (October 10, 2011): 203–16.

Basu, Avik, Jason Duvall, and Rachel Kaplan. "Attention Restoration Theory:
Exploring the Role of Soft Fascination and Mental Bandwidth." *Environment
and Behavior* 51, no. 9–10 (November 2019): 1055–81.

Baumeister, Roy F., Mark Muraven, and Dianne M. Tice. "Ego Depletion: A
Resource Model of Volition, Self-Regulation, and Controlled Processing."
Social Cognition 18, no. 2 (June 2000): 130–50.

BBC News. "US Officer Plays Taylor Swift Song to Try to Block Video." BBC. July
2, 2021. https://www.bbc.com/news/technology-57698858.

Beaney, Michael. "Russell and Frege." In *The Cambridge Companion to Bertrand
Russell*, edited by Nicholas Griffin, 128–70. Cambridge Cambridge University
Press, 2003.

Beardsley, Monroe C. "Metaphorical Senses." *Noûs* 12, no. 1 (1978): 3–16.

―――. "The Metaphorical Twist." *Philosophy and Phenomenological Research* 22,
no. 3 (1962): 293–307.

Benjamin, Ruha. *Race After Technology: Abolitionist Tools for the New Jim Code*.
Medford, MA: Polity Press, 2019.

Bhanoo, Sindya. "Who Should Translate Amanda Gorman's Work? That Question
Is Ricocheting around the Translation Industry." *Washington Post*. March 24,
2021. https://www.washingtonpost.com/entertainment/books/book-transla
tions-gorman-controversy/2021/03/24/8ea3223e-8cd5-11eb-9423-04079921c9
15_story.html.

Black, Max. *Models and Metaphors: Studies in Language and Philosophy*. Ithaca, NY:
Cornell University Press, 1962.

―――. "More About Metaphor." *Dialectica* 31, no. 34 (1977): 431–57.

Blank, Grant. "The Digital Divide Among Twitter Users and Its Implications for
Social Research." *Social Science Computer Review* 35, no. 6 (December 2017):
679–97.

Blavin, Jonathan H., and I. Glenn Cohen. "Gore, Gibson, and Goldsmith: The Evolution of Internet Metaphors in Law and Commentary." *Harvard Journal of Law and Technology* 16, no. 1 (2002): 265–87.

Bohn, Dieter. "Apple Isn't Just a Walled Garden, It's a Carrier." The Verge. June 7, 2021. https://www.theverge.com/2021/6/7/22521476/apple-walled-garden-ca rrier-app-store-innovation.

Bohrn, Isabel C., Ulrike Altmann, and Arthur M. Jacobs. "Looking at the Brains behind Figurative Language—A Quantitative Meta-Analysis of Neuroimaging Studies on Metaphor, Idiom, and Irony Processing." *Neuropsychologia* 50, no. 11 (September 2012): 2669–83.

Borgmann, Albert. *Technology and the Character of Contemporary Life: A Philosophical Inquiry*. Chicago: University of Chicago Press, 1987.

boyd, danah. *It's Complicated: The Social Lives of Networked Teens*. New Haven, CT: Yale University Press, 2014.

Brady, William J., Killian McLoughlin, Tuan N. Doan, and Molly J. Crockett. "How Social Learning Amplifies Moral Outrage Expression in Online Social Networks." *Science Advances* 7, no. 33 (August 2021). https://doi.org/10.1126/scia dv.abe5641.

Brennan, P. "Sexual Selection." *Nature Education Knowledge* 3, no. 10 (2010): 79.

Buckle, Chase. "Which Smartphone Features Really Matter to Consumers?" GWI. January 9, 2019. https://blog.gwi.com/chart-of-the-week/smartphone-featur es-consumers/.

Buckley, William F., Jr. "Do We Really Need Home Computers?" *On the Right* (syndicated column). June 19, 1982.

Burke, James. *Connections: Alternative History of Technology*. New York: Macmillan, 1980.

Cameron, Lynne. "Metaphor and Talk." In *The Cambridge Handbook of Metaphor and Thought*, edited by Raymond W. Gibbs Jr., 197–211. Cambridge: Cambridge University Press, 2008.

Carbonell, Javier, Antonio Sánchez-Esguevillas, and Belén Carro. "The Role of Metaphors in the Development of Technologies. The Case of the Artificial Intelligence." *Futures* 84 (November 2016): 145–53.

Card, Orson Scott. *Ender's Game*. New York: Tor Books, 2014.

Carr, Austin. "Will Apple's Tacky Design Philosophy Cause a Revolt?" Fast Company. September 11, 2012. https://www.fastcompany.com/1670760/will-appl es-tacky-software-design-philosophy-cause-a-revolt.

Cassirer, Ernst. *An Essay on Man*. New Haven, CT: Yale University Press, 2021.

———. "A Clue to the Nature of Man: The Symbol." In *An Essay on Man: An Introduction to a Philosophy of Human Culture*. New Haven, CT: Yale University Press, 1992.

Cawley, Conor. "10 Stats That Prove Silicon Valley Hasn't Fixed Its Diversity Problem." Tech.co. June 25, 2018. https://tech.co/news/stats-silicon-valley-fixed-diversity-problem-2018-06.

Chan, Stephanie. "Global Consumer Spending in Mobile Apps Reached a Record $111 Billion in 2020, up 30% from 2019." Sensor Tower. January 4, 2021. https://sensortower.com/blog/app-revenue-and-downloads-2020.

Charteris-Black, J. *Politicians and Rhetoric: The Persuasive Power of Metaphor*. London: Palgrave Macmillan, 2011.

Chown, Eric. "Reminiscence and Arousal: A Connectionist Model." In Proceedings of the Annual Meeting of the Cognitive Science Society, Vol. 24, 2002. https://escholarship.org/content/qt4v45r3gd/qt4v45r3gd.pdf.

———. "Spatial Prototypes." In Spatial Behavior and Linguistic Representation, edited by Thora Thenbrink, Jan Wiener, and Christophe Claramunt, 97–114. Oxford: Oxford University Press, 2013.

Chown, Eric, and Fernando Nascimento. "Software and Metaphors: The Hermeneutical Dimensions of Software Development." In *Interpreting Technology*, edited by Wessel Reijers, Alberto Romele, and Mark Coeckelbergh, 229–48. Lanham, MD: Rowman & Littlefield, 2021.

Cimprich, Bernadine. "Development of an Intervention to Restore Attention in Cancer Patients." *Cancer Nursing* 16, no. 2 (April 1993): 83–92.

Coeckelbergh, Mark. *AI Ethics*. Cambridge, MA: MIT Press, 2020.

———. "Language and Technology: Maps, Bridges, and Pathways." *AI & Society* 32, no. 2 (May 2017): 175–89.

Coeckelbergh, Mark, and Wessel Reijers. "Narrative Technologies: A Philosophical Investigation of the Narrative Capacities of Technologies by Using Ricoeur's Narrative Theory." *Human Studies* 39, no. 3 (September 2016): 325–46.

Cohen, Jean. "Structure Du Langage Poétique." *Les Etudes Philosophiques* 22, no. 2 (1967): 219.

Collins, Ben, and Brandy Zadrozny. "Anti-Vaccine Groups Changing into 'Dance Parties' on Facebook to Avoid Detection." NBC News. July 22, 2021. https://www.nbcnews.com/tech/tech-news/anti-vaccine-groups-changing-dance-parties-facebook-avoid-detection-rcna1480.

MacMillan Dictionary. "Common Metaphors." *MacMillan Dictionary* (blog). May 11, 2011. https://www.macmillandictionaryblog.com/common-metaphors.

Craik, Kenneth J. "The Nature of Explanation." *Journal of Philosophy* 40, no. 24 (1943): 667–69. https://doi.org/10.2307/2018933.

Cridland, James. "'Follow Our Podcast': Apple Podcasts to Stop Using 'Subscribe.'" Podnews. March 9, 2021. https://podnews.net/update/follow-not-subscribe.

Damasio, Antonio R. *Descartes' Error: Emotion, Reason and the Human Brain*. New York: Random House, 2006.

Daniel, P. "Four Years of Disruption: Cell Phone Industry Financials 2007-2011." PhoneArena. July 14, 2011. https://www.phonearena.com/news/Four-years-of -disruption-cell-phone-industry-financials-2007-2011_id20153.

Demjaha, Albese, Jonathan Spring, Ingolf Becker, Simon Parkin, and Angela Sasse. "Metaphors Considered Harmful? An Exploratory Study of the Effectiveness of Functional Metaphors for End-to-End Encryption." In *Proceedings of 2018 Workshop on Usable Security (USEC)*. Reston, VA: Internet Society, 2018. https://doi.org/10.14722/usec.2018.23015.

Desjardins, Jeff. "How Long Does It Take to Hit 50 Million Users." Visual Capitalist. 2018. https://www.visualcapitalist.com/how-long-does-it-take-to-hit -50-million-users/.

Doctorow, Cory. "The Coming Civil War over General Purpose Computing." Boing Boing. August 23, 2012. https://boingboing.net/2012/08/23/civilwar .html.

Erickson, Christine. "A Brief History of Text Messaging." Mashable. September 21, 2012. http://mashable. com/2012/09/21/text-Messaging-History.

Erickson, Thomas D. "Working with Interface Metaphors." In *Readings in Human– Computer Interaction: Toward the Year 2000*. 2nd ed., edited by Ronald M. Baecker, Jonathan Grudin, William A. S. Buxton, and Saul Greenberg, 147–51 (San Francisco: Morgan Kaufmann, 1995). https://doi.org/10.1016/b978-0-08-0515 74-8.50018-2.

Ervas, Francesca, Elisabetta Gola, and Maria Grazia Rossi, eds. *Metaphor in Communication, Science and Education*. Berlin: de Gruyter, 2017.

Evans, Benedict. "The Death of the Newsfeed." *Benedict Evans* (blog). April 3, 2018. https://www.ben-evans.com/benedictevans/2018/4/2/the-death-of-the-news feed.

Eviatar, Zohar, and Marcel Adam Just. "Brain Correlates of Discourse Processing: An fMRI Investigation of Irony and Conventional Metaphor Comprehension." *Neuropsychologia* 44, no. 12 (June 23, 2006): 2348–59.

ExplainingComputers. "5 Most Annoying Computing Things!" June 20, 2021. Video, 13:15. https://www.youtube.com/watch?v=HuF4NfUok3Q.

Eyal, Nir. "The *New York Times* Uses the Very Dark Patterns It Derides." *Nirandfar* (blog). May 30, 2021. https://www.nirandfar.com/cancel-new-york-times/.

Facebook. "A/B Tests Types Available on Facebook." Meta. Accessed July 30, 2021. https://www.facebook.com/business/help/1159714227408868?content_id=h m7zSCLtFLb30Qs&ref=sem_smb&utm_source=GOOGLE&utm_medium= fbsmbsem&utm_campaign=PFX_SEM_G_BusinessAds_US_EN_Brand_Ex act_Desktop&utm_content=Ads-Testing_Evaluating&kenid=_k_CjwKCAjw xo6lBhBKEiwAXSYBs4qIWko1GKG8bxS7msQc6u48GzSGbwnY7dLvhRItLU m6TM5KWvCjrhoCtioQAvD_BwE_k_&utm_term=facebook%20a%2Fb%20

testing&utm_ct=EVG&gclid=CjwKCAjwxo6IBhBKEiwAXSYBs4qIWko1GKG
8bxS7msQc6u48GzSGbwnY7dLvhRItLUm6TM5KWvCjrhoCtioQAvD_BwE.

Fiegerman, Seth. "The Experts Speak: Here's What People Predicted Would Happen When The iPhone Came Out." Business Insider. June 29, 2012. https://
www.businessinsider.com/iphone-predictions-from-2007-2012-6.

Floridi, Luciano. The Fourth Revolution: How the Infosphere Is Reshaping Human
Reality. Oxford: Oxford University Press, 2014.

Forman, Paul. "The Primacy of Science in Modernity, of Technology in Postmodernity, and of Ideology in the History of Technology." History and Technology
23, no. 1–2 (March 2007): 1–152.

Frege, Gottlob. "Sense and Reference." The Philosophical Review 57, no. 3 (1948):
209–30.

Frost, Robert. "Education by Poetry." In The Selected Prose of Robert Frost, edited
by Hyde Cox and Edward Connery Lathem, 33–46. New York: Holt, Rinehart
and Winston, 1966.

Gadamer, Hans-Georg. Truth and Method. New York: Crossroad, 1989.

Gatsou, Chrysoula. "The Importance of Metaphors for User Interaction with
Mobile Devices." In Design, User Experience, and Usability: Users and Interactions, edited by Aaron Marcus, 520–29. Cham: Springer International Publishing, 2015.

Gayomali, Chris. "The Text Message Turns 20: A Brief History of SMS." The
Weekly Law Reports. 2012. https://theweek.com/articles/469869/text-mess
age-turns-20-brief-history-sms.

Geary, James. I Is an Other: The Secret Life of Metaphor and How It Shapes the Way
We See the World. New York: Harper Collins, 2011.

Gibbs, Raymond W., Jr., ed. The Cambridge Handbook of Metaphor and Thought.
Cambridge: Cambridge University Press, 2008.

Goatly, Andrew. Washing the Brain: Metaphor and Hidden Ideology. Amsterdam:
John Benjamins Publishing, 2007.

Goldhaber, Michael H. "The Attention Economy and the Net." First Monday 2, no.
4 (April 7, 1997). https://doi.org/10.5210/fm.v2i4.519.

Goode, Lauren. "I Called Off My Wedding. The Internet Will Never Forget."
Wired. April 6, 2021. https://www.wired.com/story/weddings-social-media
-apps-photos-memories-miscarriage-problem/.

Gorman, Amanda (@TheAmandaGorman). Twitter Timeline. Accessed March 7,
2021. https://twitter.com/TheAmandaGorman.

Gould, Skye. "It Took 75 Years for the Telephone to Reach 100 Million Users and
It Took Candy Crush 15 Months." Business Insider. July 28, 2015. https://www
.businessinsider.com/it-took-75-years-for-the-telephone-to-reach-100-milli
on-users-and-it-took-candy-crush-15-months.

Gramlich, John. "10 Facts about Americans and Facebook." Pew Research. June 1, 2021. https://www.pewresearch.org/fact-tank/2021/06/01/facts-about-amer icans-and-facebook/.

Greene, Melissa Fay. "You Won't Remember the Pandemic the Way You Think You Will." *The Atlantic*, April 6, 2021. https://www.theatlantic.com/magazine/arch ive/2021/05/how-will-we-remember-covid-19-pandemic/618397/.

Gruber, John. "Theme Parks and Public Parks." Daring Fireball. June 6, 2021. https://daringfireball.net/linked/2021/06/06/theme-parks-and-public-parks.

Harmon, Sarah, and Katie McDonough. "The Draw-A-Computational-Creativity-Researcher Test (DACCRT): Exploring Stereotypic Images and Descriptions of Computational Creativity." In *Proceedings of the 10th International Conference on Computational Creativity*, edited by Kazjon Grace, Michael Cook, Dan Ventura, and Mary Lou Maher, 243–49. Charlotte, NC: Association for Computational Creativity, 2019.

Hawkins, Jeff. *A Thousand Brains: A New Theory of Intelligence*. New York: Basic Books, 2021.

Hayles, N. Katherine. Writing Machines. Cambridge, MA: MIT Press, 2002.

Heath, Alex. "Facebook Plans First Smartwatch for Next Summer with Two Cameras, Heart Rate Monitor." The Verge. June 9, 2021. https://www.theverge.com /2021/6/9/22526266/facebook-smartwatch-two-cameras-heart-rate-monitor.

Hebb, Donald O. *The Organization of Behavior*. New York: Wiley, 1949.

He, Ray C. "Introducing New Like and Share Buttons." *Facebook for Developers* (blog). November 6, 2013. https://developers.facebook.com/blog/post/2013/11 /06/introducing-new-like-and-share-buttons/.

Hobbes, Thomas. *Leviathan*. Edited by J. C. A. Gaskin. Oxford: Oxford University Press, 1998.

Holyoak, Keith J., and Dušan Stamenković. "Metaphor Comprehension: A Critical Review of Theories and Evidence." *Psychological Bulletin* 144, no. 6 (June 2018): 641–71.

Hutchinson, Thomas H. *Here Is Television, Your Window to the World*. New York: Hastings House, 1950.

Hwang, Tim, and Karen Levy. "'The Cloud' and Other Dangerous Metaphors." *The Atlantic*, January 20, 2015. https://www.theatlantic.com/technology/arch ive/2015/01/the-cloud-and-other-dangerous-metaphors/384518/.

Ihde, Don. *Philosophy of Technology: An Introduction*. St. Paul, MN: Paragon House, 1998.

———. *Technology and the Lifeworld: From Garden to Earth*. Bloomington: Indiana University Press, 1990.

Ivancich, Eric J. "Fostering Clarity to Achieve Reasonableness." In *Fostering Reasonableness: Supportive Environments for Bringing Out Our Best*, edited by

Rachel Kaplan and Avik Basu. Ann Arbor, MI: Maize Publishing, 2015. https://doi.org/10.3998/maize.13545970.0001.001.

Jahnke, Marcus. "Revisiting Design as a Hermeneutic Practice: An Investigation of Paul Ricoeur's Critical Hermeneutics." *Design Issues* 28, no. 2 (2012): 30–40.

James, William. *Psychology: The Briefer Course.* Edited by Gordon Allport. New York: Harper & Row, 1961.

Jenner, Leontine. "Backdoor: How a Metaphor Turns into a Weapon." *Digital Society Blog.* Alexander von Humboldt Institute for Internet and Society. 2018. https://www.hiig.de/en/backdoor-how-a-metaphor-turns-into-a-weapon/.

Jin, Huifeng. "Influence of Icon Design Style on User's Cognition." In *Proceedings of the 6th International Conference on Humanities and Social Science Research* (ICHSSR 2020), edited by Xuemei Du, Chunyan Huang, and Yulin Zhong, 550–53. Paris: Atlantis Press, 2020.

Johnson, Khari. "This New Way to Train AI Could Curb Online Harassment." Wired, August 4, 2021. https://www.wired.com/story/new-way-train-ai-curb-online-harassment/.

Johnson, Mark. *The Body in the Mind: The Bodily Basis of Meaning, Imagination, and Reason.* Chicago: University of Chicago Press, 1987.

Jones, Nicola. "How to Stop Data Centres from Gobbling up the World's Electricity." *Nature* 561, no. 7722 (September 2018): 163–66.

Just, Marcel Adam, Vladimir L. Cherkassky, Sandesh Aryal, and Tom M. Mitchell. "A Neurosemantic Theory of Concrete Noun Representation Based on the Underlying Brain Codes." *PLOS One* 5, no. 1 (January 13, 2010): article e8622.

Kahneman, Daniel. *Thinking, Fast and Slow.* New York: Farrar, Strauss and Giroux, 2011.

Kahneman, Daniel, Olivier Sibony, and Cass R. Sunstein. *Noise: A Flaw in Human Judgment.* New York: Little, Brown, 2021.

Kaplan, Rachel. "Attention! That's a Precious Resource." *Our Planet* 2017, no. 2 (May 2018): 28–29. https://doi.org/10.18356/63bb4c1f-en.

Kaplan, Stephen. "Attention and Fascination: The Search for Cognitive Clarity." In *Humanscape: Environments for People,* edited by Stephen Kaplan and Rachel Kaplan, 84–93. Belmont, CA: Duxbury, 1978.

———. "Beyond Rationality: Clarity-Based Decision Making." In *Environment, Cognition, and Action: An Integrated Approach,* edited by Thomas Garling and Gary W. Evans, 171–90. Oxford: Oxford University Press, 1991.

———. "The Expertise Challenge." In *Fostering Reasonableness: Supportive Environments for Bringing Out Our Best,* edited by Rachel Kaplan and Avik Basu, 43–53. Ann Arbor, MI: Maize Books, 2015. https://deepblue.lib.umich.edu/handle/2027.42/150716.

Kaplan, Stephen, and Marc G. Berman. "Directed Attention as a Common Resource for Executive Functioning and Self-Regulation." *Perspectives on Psychological Science* 5, no. 1 (January 2010): 43–57.

Katzenbach, Christian, and Stefan Larsson. "How Metaphors Shape the Digital Society." *Digital Society Blog*. Alexander von Humboldt Institute for Internet and Society. 2018. https://www.hiig.de/en/dossier/how-metaphors-shape-the-digital-society/.

Kemp, Simon. "Digital 2021: Global Overview Report." Datareportal. January 27, 2021. https://datareportal.com/reports/digital-2021-global-overview-report

Kingsley-Hughes, Adrian. "Who Do I Pay to Get the 'Phone' Removed from My iPhone?" ZDNet. April 12, 2021. https://www.zdnet.com/article/who-do-i-pay-to-get-the-phone-removed-from-my-iphone/.

Kintsch, Walter. "Metaphor Comprehension: A Computational Theory." *Psychonomic Bulletin & Review* 7, no. 2 (June 2000): 257–66.

Kirby, John T. "Aristotle on Metaphor." *American Journal of Philology* 118, no. 4 (1997): 517–54.

Kleinsmith, Lewis J., and Stephen Kaplan. "Interaction of Arousal and Recall Interval in Nonsense Syllable Paired-Associate Learning." *Journal of Experimental Psychology* 67, no. 2 (February 1964): 124–26.

———. "Paired-Associate Learning as a Function of Arousal and Interpolated Interval." *Journal of Experimental Psychology* 65, no. 2 (February 1963): 190–93.

Kramer, Adam D. I., Jamie E. Guillory, and Jeffrey T. Hancock. "Experimental Evidence of Massive-Scale Emotional Contagion through Social Networks." *Proceedings of the National Academy of Sciences of the United States of America* 111, no. 24 (June 17, 2014): 8788–90.

Krishna, Aradhna. "How Did 'White' Become a Metaphor for All Things Good?" The Conversation. July 6, 2020. http://theconversation.com/how-did-white-become-a-metaphor-for-all-things-good-140674.

Kwai, Isabella. "Consumer Groups Target Amazon Prime's Cancellation Process." *New York Times*, January 14, 2021. https://www.nytimes.com/2021/01/14/world/europe/amazon-prime-cancellation-complaint.html.

Lachaud, Christian. "Conceptual Metaphors and Embodied Cognition: EEG Coherence Reveals Brain Activity Differences between Primary and Complex Conceptual Metaphors during Comprehension." *Cognitive Systems Research* 22–23 (June 2013): 12–26.

Lakoff, George, and Mark Johnson. *Metaphors We Live By*. Chicago: University of Chicago Press, 1980.

Landau, Mark J., Brian P. Meier, and Lucas A. Keefer. "A Metaphor-Enriched Social Cognition." *Psychological Bulletin* 136, no. 6 (November 2010): 1045–67.

Latour, Bruno. "On Technical Mediation." *Common Knowledge* 3, no. 2 (1994): 29–64.

Leonardi, Paul M. "Digital Materiality? How Artifacts without Matter, Matter." *First Monday* 15, no. 6 (June 2010). https://doi.org/10.5210/fm.v15i6.3036.

Levy, Adam. "Investing in FAANG or MAMAA Stocks." The Motley Fool. Updated July 21, 2022. https://www.fool.com/investing/stock-market/market-sectors/information-technology/faang-stocks/.

Lewis, David. "General Semantics." In *Montague Grammar*, edited by Barbara H. Partee, 1–50. New York: Academic Press, 1976.

Ling, Richard. *Taken for Grantedness: The Embedding of Mobile Communication into Society*. Cambridge, MA: MIT Press, 2012.

Lomas, Natasha. "Targeted Ads Offer Little Extra Value for Online Publishers, Study Suggests." TechCrunch. May 31, 2019. http://techcrunch.com/2019/05/31/targeted-ads-offer-little-extra-value-for-online-publishers-study-suggests/.

Lorch, Mark. "How Children Are Spoofing Covid-19 Tests with Soft Drinks." BBC. July 6, 2021. https://www.bbc.com/future/article/20210705-how-children-are-spoofing-covid-19-tests-with-soft-drinks.

Low, Graham, Zazie Todd, Alice Deignan, and Lynne Cameron, eds. *Researching and Applying Metaphor in the Real World*. Amsterdam: John Benjamins Publishing, 2010.

Lycan, William G. *Philosophy of Language: A Contemporary Introduction*. 3rd ed. London: Routledge, 2018.

Mahdawi, Arwa. "Targeted Ads Are One of the World's Most Destructive Trends. Here's Why." *The Guardian*, November 5, 2019. http://www.theguardian.com/world/2019/nov/05/targeted-ads-fake-news-clickbait-surveillance-capitalism-data-mining-democracy.

Mandler, George. "Consciousness: Respectable, Useful, and Probably Necessary." In *Information Processing and Cognition*, edited by Robert L. Solso, 229–54. Hillsdale, NJ: Lawrence Erlbaum, 1975.

Marakas, George M., Richard D. Johnson, and Jonathan W. Palmer. "A Theoretical Model of Differential Social Attributions toward Computing Technology: When the Metaphor Becomes the Model." *International Journal of Human-Computer Studies* 52, no. 4 (April 2000): 719–50.

Markham, Annette. "The Limits of the Imaginary: Challenges to Intervening in Future Speculations of Memory, Data, and Algorithms." *New Media & Society* 23, no. 2 (February 2021): 382–405.

Marshall, Alex. "Amanda Gorman's Poetry United Critics. It's Dividing Translators." *New York Times*, March 26, 2021. https://www.nytimes.com/2021/03/26/books/amanda-gorman-hill-we-climb-translation.html.

McCloud, Scott. "Google Chrome." Google. 2008. https://www.google.com/goo glebooks/chrome/.

McFadden George "Interpretation Theory: Discourse and the Surplus of Meaning." *Journal of Aesthetics and Art Criticism* 36, no. 3 (1976): 365–67.

Mead, Margaret, and Rhoda Metraux. "Image of the Scientist among High-School Students." *Science* 126, no. 3270 (1957): 384–90.

Meconis, Dylan, Scott McCloud, and Syne Mitchell. "Learning Machine Learning." Google. 2019. https://cloud.google.com/products/ai/ml-comic-1.

Michel, Johann. *Homo Interpretans: Towards a Transformation of Hermeneutics.* Translated by David Pellauer. Lanham, MD: Rowman & Littlefield, 2019.

Milian, Mark. "Why Text Messages Are Limited to 160 Characters." *Los Angeles Times Blog.* May 3, 2009. https://latimesblogs.latimes.com/technology/2009 /05/invented-text-messaging.html.

Miller, Bryan. "How Mobile Technology Is Changing the Way We Dine Out." WSJ Online, October 25, 2013. https://www.wsj.com/articles/SB100014240527023 03680404579143903633457212.

Miller, Daniel, Elisabetta Costa, Nell Haynes, Tom McDonald, Razvan Nicolescu, Jolynna Sinanan, Juliano Spyer, Shriram Venkatraman, and Xinyuan Wang. "Online and Offline Relationships." In *How the World Changed Social Media,* 100–113. London: UCL Press, 2016.

Mitcham, Carl. *Thinking Through Technology: The Path Between Engineering and Philosophy.* Chicago: University of Chicago Press, 1994.

Mohsin, Maryam. "10 TikTok Statistics that You Need to Know in 2020." Oberlo. February 16, 2021. https://www.oberlo.com/blog/tiktok-statistics.

Monkman, Betty C. "Pardoning the Thanksgiving Turkey." White House Historical Association. Accessed August 3, 2021. https://www.whitehousehistory.org /pardoning-the-thanksgiving-turkey.

Monteiro, Mike. *Ruined by Design: How Designers Destroyed the World, and What We Can Do to Fix It.* Independently published, 2019.

Moran, Thomas P., and Shumin Zhui. "Beyond the Desktop Metaphor in Seven Dimensions." In *Designing Integrated Digital Work Environments: Beyond the Desktop Metaphor,* edited by Victor Kaptelinin and Mary Czerwinski, 335–54. Cambridge, MA: MIT Press, 2007.

Morris, Margaret E. *Left to Our Own Devices: Outsmarting Smart Technology to Reclaim Our Relationships, Health, and Focus.* Cambridge, MA: MIT Press, 2018.

Morris, Michael W., Oliver J. Sheldon, Daniel R. Ames, and Maia J. Young. "Metaphors and the Market: Consequences and Preconditions of Agent and Object Metaphors in Stock Market Commentary." *Organizational Behavior and Human Decision Processes* 102, no. 2 (March 2007): 174–92.

Munroe, Randall. *Thing Explainer: Complicated Stuff in Simple Words.* London: Hachette UK, 2015.

Musolff, Andreas. *Metaphor and Political Discourse: Analogical Reasoning in Debates about Europe.* Basingstoke: Palgrave Macmillan, 2004.

Neely, Amber. "TikTok Can Collect Users' Biometric Data According to Its New Privacy Policy." AppleInsider. June 4, 2021. https://appleinsider.com/articles /21/06/04/tiktok-can-collect-users-biometric-data-according-to-its-new-pri vacy-policy.

Neisser, Ulric, and Nicole Harsch. "Phantom Flashbulbs: False Recollections of Hearing the News about Challenger." *Affect and Accuracy in Recall: Studies of "Flashbulb" Memories,* edited by Eugene Winograd and Ulric Neisser, 9–31. Cambridge: Cambridge University Press, 1992.

Noble, Safiya Umoja. *Algorithms of Oppression: How Search Engines Reinforce Racism.* New York: New York University Press, 2018.

Norman, Don. *The Design of Everyday Things,* revised and expanded ed. New York: Basic Books, 2013.

Oades, Robert D. "The Role of Noradrenaline in Tuning and Dopamine in Switching between Signals in the CNS." *Neuroscience and Biobehavioral Reviews* 9, no. 2 (June 1985): 261–82.

O'Dea, S. "Mobile Device Daily Usage Time in U.S. 2014-2021." Statista. February 27, 2020. https://www.statista.com/statistics/1045353/mobile-device-daily-us age-time-in-the-us/.

O'Neil, Cathy. *Weapons of Math Destruction: How Big Data Increases Inequality and Threatens Democracy.* New York: Crown, 2016.

O'Reilly, Tim. "Data Is the New Sand." The Information. February 24, 2021. https://www.theinformation.com/articles/data-is-the-new-sand?utm_sourc e=ti_app.

Oster, Judith. "Frost's Poetry of Metaphor." In *The Cambridge Companion to Robert Frost,* edited by Robert Fagen, 155–78. Cambridge: Cambridge University Press, 2001. https://doi.org/10.1017/ccol052163248x.008.

Oswald, David. "Affordances and Metaphors Revisited: Testing Flat vs. Skeuomorph Design with Digital Natives and Digital Immigrants." In *Proceedings of the 32nd International BCS Human Computer Interaction Conference (HCI),* 1–11. Swindon, UK: BCS Learning & Development, 2018.

PairPlayApp. "PairPlay." Accessed August 17, 2021. https://www.pairplayapp.com.

Palermo, Elizabeth. "Who Invented the Lightbulb?" Live Science, November 23, 2021. https://www.livescience.com/43424-who-invented-the-light-bulb.html.

Payne, Keith, Laura Niemi, and John M. Doris. "How to Think about 'Implicit Bias.'" *Scientific American,* March 27, 2018. https://www.scientificamerican .com/article/how-to-think-about-implicit-bias/.

Perez, Caroline Criado. *Invisible Women: Data Bias in a World Designed for Men.* New York: Abrams, 2019.

Phil, B. "Importance of First – Why Nokia Cares about Super Affordable Mobile Phones." *Nokia Conversations: The Official Nokia Blog.* September 16, 2013. https://web.archive.org/web/20130916190754/http://conversations.nokia .com/2013/09/13/importance-of-first-why-nokia-cares-about-super-affordab le-mobile-phones.

Pierce, Jessica. "What the Grieving Mother Orca Tells Us about How Animals Experience Death." The Conversation. August 24, 2018. http://theconversa tion.com/what-the-grieving-mother-orca-tells-us-about-how-animals-expe rience-death-101230.

Popper, Nathaniel. "As Diners Flock to Delivery Apps, Restaurants Fear for Their Future." *New York Times,* June 9, 2020. https://www.nytimes.com/2020/06/09 /technology/delivery-apps-restaurants-fees-virus.html.

Postman, Neil. *Amusing Ourselves to Death: Public Discourse in the Age of Show Business.* New York: Penguin, 1985.

"Project Implicit." Harvard University. Accessed August 5, 2021. https://implicit .harvard.edu/implicit/index.jsp.

Rafalow, Matthew H., and Britni L. Adams. "Digitally Mediated Connections and Relationship Persistence in Bar Settings." *Symbolic Interaction* 40, no. 1 (2017): 25–42.

Ranzini, Giulia, and Christoph Lutz. "Love at First Swipe? Explaining Tinder Self-presentation and Motives." *Mobile Media & Communication* 5, no. 1 (2017): 80–101.

Reese, Elaine, and Robyn Fivush. "The Development of Collective Remembering." *Memory* 16, no. 3 (April 2008): 201–12.

Richards, Ivor Armstrong. *The Philosophy of Rhetoric.* Oxford: Oxford University Press, 1936.

———. *Principles of Literary Criticism.* London: Routledge, 2017.

Ricoeur, Paul. *From Text to Action: Essays in Hermeneutics II.* Translated by John B. Thompson and Kathleen Blamey. New York: Bloomsbury Publishing, 2008.

———. *Hermeneutics and the Human Sciences: Essays on Language, Action and Interpretation.* Cambridge: Cambridge University Press, 1981.

———. *Interpretation Theory: Discourse and the Surplus of Meaning.* Fort Worth: Texas Christian University Press, 1976.

———. "Metaphor and the Central Problem of Hermeneutics." *Graduate Faculty Philosophy Journal* 3, no. 1 (April 1, 1973): 42–58.

———. "The Metaphorical Process as Cognition, Imagination, and Feeling." *Critical Inquiry* 5, no. 1 (October 1978): 143–59.

———. "The Power of Speech: Science and Poetry." *Philosophy Today* 29, no. 1 (February 1985): 59–70.

———. *The Rule of Metaphor: The Creation of Meaning in Language.* London: Psychology Press, 2003.

———. *The Rule of Metaphor: Multi-Disciplinary Studies of the Creation of Meaning in Language.* Toronto: University of Toronto Press, 1978.

———. *Time and Narrative.* Vol. 1. Chicago: University of Chicago Press, 1984.

Roose, Kevin. "Why Are We Still Calling the Things in Our Pockets 'Cell Phones'?" Intelligencer. June 24, 2014. https://nymag.com/intelligencer/2014/06/why -are-we-still-calling-them-cell-phones.html.

Rosales, Andrea, and Mireia Fernández-Ardèvol. "Ageism in the Era of Digital Platforms." *Convergence* 26, no. 5–6 (December 2020): 1074–87.

Rosch, Eleanor. "Principles of Categorization." In *Concepts: Core Readings*, edited by Eric Margolis and Stephen Laurence, 251–70. Cambridge, MA: MIT Press, 1999.

Rowell, David. "Dead Musicians Are Taking the Stage Again in Hologram Form. Is This the Kind of Encore We Really Want?" *Washington Post*, October 30, 2019. https://www.washingtonpost.com/magazine/2019/10/30/dead-musicians-are -taking-stage-again-hologram-form-is-this-kind-encore-we-really-want/.

Sanders, Carol. "Introduction: Saussure Today." In *The Cambridge Companion to Saussure*, edited by Carol Sanders, 1–6. Cambridge: Cambridge University Press, 2004. https://doi.org/10.1017/CCOL052180051X.001.

Sap, Maarten, Dallas Card, Saadia Gabriel, Yejin Choi, and Noah A. Smith. "The Risk of Racial Bias in Hate Speech Detection." In *Proceedings of the 57th Annual Meeting of the Association for Computational Linguistics*, edited by Preslav Nakov and Alexis Palmer, 1668–78. Florence, Italy: Association for Computational Linguistics, 2019.

Saussure, Ferdinand Mongin de. *Course in General Linguistics.* New York: Columbia University Press, 2011.

Schatzberg, Eric. *Technology: Critical History of a Concept.* Chicago: University of Chicago Press, 2018.

Schön, Donald A. "Generative Metaphor: A Perspective on Problem-Setting in Social Policy." In *Metaphor and Thought*, 2nd ed., edited by Andrew Ortony, 135–61. Cambridge: Cambridge University Press, 1993.

Schwartz, Oscar. "In 2016, Microsoft's Racist Chatbot Revealed the Dangers of Online Conversation." IEEE Spectrum. November 25, 2019. https://spectrum .ieee.org/in-2016-microsofts-racist-chatbot-revealed-the-dangers-of-online -conversation.

Searle, John R. *Speech Acts: An Essay in the Philosophy of Language.* Cambridge: Cambridge University Press, 1969.

Shakespeare, William. *As You Like It*. Edited by Alan Brissenden. Oxford: Oxford University Press, 1993. https://doi.org/10.1093/oseo/instance.00005809.

———. *Troilus and Cressida*. Edited by Barbara A. Mowat and Paul Werstine. Washington, DC: Folger Shakespeare Library, 2007.

Sheffi, Yossi. "A Failure to Treat Workers with Respect Could Be Uber's Achilles' Heel." MIT Technology Review. September 22, 2014. https://www.technolog yreview.com/2014/09/22/12491/a-failure-to-treat-workers-with-respect-cou ld-be-ubers-achilles-heel/.

Shin, Jiwoong, and Jungju Yu. "Targeted Advertising and Consumer Inference." SSRN. September 7, 2020. https://doi.org/10.2139/ssrn.3688258.

Simanowski, Roberto. *Facebook Society: Losing Ourselves in Sharing Ourselves*. New York: Columbia University Press, 2018.

Simon, Herbert A. "Designing Organizations for an Information-Rich World." In *Computers, Communications, and the Public Interest*, edited by M. Greenberger, 37–52. Baltimore, MD: John Hopkins University Press, 1971.

Simonite, Tom. "Facebook's Like Buttons Will Soon Track Your Web Browsing to Target Ads." MIT Technology Review. September 16, 2015. https://www.tec hnologyreview.com/2015/09/16/166222/facebooks-like-buttons-will-soon-tr ack-your-web-browsing-to-target-ads/.

Sismondo, Sergio. *Science without Myth: On Constructions, Reality, and Social Knowledge*. Albany: State University of New York Press, 1996.

Slack, Jennifer Daryl, and J. Macgregor Wise. *Culture and Technology: A Primer*. 2nd ed. New York: Peter Lang Publishing, 2005.

Slepian, Michael L., Max Weisbuch, Abraham M. Rutchick, Leonard S. Newman, and Nalini Ambady. "Shedding Light on Insight: Priming Bright Ideas." *Journal of Experimental Social Psychology* 46, no. 4 (July 2010): 696–700.

Sporns, Olaf, Giulio Tononi, and Rolf Kötter. "The Human Connectome: A Structural Description of the Human Brain." *PLoS Computational Biology* 1, no. 4 (2005): e42. https://doi.org/10.1371/journal.pcbi.0010042.

Steen, Gerard J., Aletta G. Dorst, J. Berenike Herrmann, Anna A. Kaal, and Tina Krennmayr. "Metaphor in Usage." *Cognitive Linguistics* 21, no. 4 (November 2010): 765–96.

Stern, Joanna. Investigation: How TikTok's Algorithm Figures out Your Deepest Desires. *Wall Street Journal*, July 21, 2021. https://on.wsj.com/3hR6GzA.

———. "iPhone? AirPods? MacBook? You Live in Apple's World. Here's What You Are Missing." *Wall Street Journal*, June 4, 2021. https://www.wsj.com/articles /iphone-airpods-macbook-you-live-in-apples-world-heres-what-you-are-mi ssing-11622817653.

Strawson, P. F. "On Referring." *Mind* 59, no. 235 (July 1950): 320–44.

———. "VI—Critical Notice." *Mind* 63, no. 249 (January 1954): 70–99.

Tendahl, Markus, and Raymond W. Gibbs. "Complementary Perspectives on Metaphor: Cognitive Linguistics and Relevance Theory." *Journal of Pragmatics* 40, no. 11 (November 2008): 1823–64.

Tenner, Edward. *Our Own Devices: How Technology Remakes Humanity.* New York: Vintage Books, 2004.

———. *Why Things Bite Back: Technology and the Revenge of Unintended Consequences.* New York: Vintage Books, 1997.

"Terminology." Junto. Accessed August 5, 2022. https://learn.junto.foundation /the-app/terminology.

Thompson, Stuart A., and Charlie Warzel. "Opinion: How Facebook Became a Tool of the Far Right." *New York Times*, January 14, 2021. https://www.nytim es.com/2021/01/14/opinion/facebook-far-right.html.

Tong, Wenley, Sebastian Gold, Samuel Gichohi, Mihai Roman, and Jonathan Frankle. "Why King George III Can Encrypt." 2014. https://www.cs.princet on.edu/~arvindn/teaching/spring-2014-privacy-technologies/king-george-iii -encrypt.pdf.

Turkle, Sherry. *Alone Together: Why We Expect More from Technology and Less from Each Other.* London: Hachette UK, 2017.

———. *Reclaiming Conversation: The Power of Talk in a Digital Age.* New York: Penguin, 2016.

Turk, Victoria. "In a Touch-Free World, the QR Code Is Having Its Moment." *Wired.* August 18, 2020. https://www.wired.com/story/in-a-touch-free-wor ld-the-qr-code-is-having-its-moment/.

"United States: Online Dating Users in the U.S. 2017–2024." Statista. July 5, 2021. https://www.statista.com/statistics/417654/us-online-dating-user-numbers/.

van den Boomen, Marianne V. T. *Transcoding the Digital: How Metaphors Matter in New Media.* Amsterdam: Institute of Network Cultures, 2014.

van Leeuwen, Edwin J. C., Katherine A. Cronin, and Daniel B. M. Haun. "Tool Use for Corpse Cleaning in Chimpanzees." *Scientific Reports* 7 (March 2017): article 44091.

Varona, Paola de. "5 Experts Explain mRNA Vaccines for Non-Science People." Verywell Health. December 21, 2020. https://www.verywellhealth.com/expla ining-mrna-vaccines-experts-social-media-5092888.

Vilgis, Thomas A. "Evolution—Culinary Culture—Cooking Technology." In *Culinary Turn: Aesthetic Practice of Cookery*, edited by Nicolaj van der Meulen and Jörg Wiesel, 149–60. Bielefeld: Transcript Verlag, 2017.

Wachter-Boettcher, Sara. *Technically Wrong: Sexist Apps, Biased Algorithms, and Other Threats of Toxic Tech.* New York: W. W. Norton, 2017.

Wakefield, Jane. "People Devote Third of Waking Time to Mobile Apps." BBC. January 12, 2022. https://www.bbc.com/news/technology-59952557.

"Watch & Manage Your Memories." Google Photos Help. Google. Accessed August 6, 2021. https://support.google.com/photos/answer/9454489?co=GENIE.Platform%3DAndroid&hl=en.

Webster, Tom. "Subscription Confusion, and the State of Podcasting Data." *I Hear Things* (blog). February 26, 2021. https://tomwebster.substack.com/p/subscription-confusion-and-the-state?utm_source=podnews.net&utm_medium=web&utm_campaign=podnews.net:2021-03-09.

"Welcome to Junto." Junto. Accessed August 5, 2022. https://learn.junto.foundation/.

Westover, Tara. *Educated: A Memoir*. New York: Random House, 2018.

Whitten, Alma, and J. D. Tygar. "Why Johnny Can't Encrypt: A Usability Evaluation of PGP 5.0." In *Security and Usability: Designing Secure Systems That People Can Use*, edited by Lorrie Faith Cranor and Simson Garfinkel, 679–702. Sebastopol, CA: O'Reilly, 2005.

Williams, Lawrence E., and John A. Bargh. "Experiencing Physical Warmth Promotes Interpersonal Warmth." *Science* 322, no. 5901 (October 24, 2008): 606–7.

———. "Keeping One's Distance: The Influence of Spatial Distance Cues on Affect and Evaluation." *Psychological Science* 19, no. 3 (March 2008): 302–8.

Wittgenstein, Ludwig. *Philosophical Investigations*. 4th ed., edited by P. M. S. Hacker and Joachim Schulte. Translated by G. E. M. Anscombe, P. M. S. Hacker, and Joachim Shulte. Malden, MA: Wiley-Blackwell, 2009.

Woo, Erin. "QR Codes Are Here to Stay. So Is the Tracking They Allow." *New York Times*, July 26, 2021. https://www.nytimes.com/2021/07/26/technology/qr-codes-tracking.html.

Wright, Mic. "The Original iPhone Announcement Annotated." The Next Web. September 9, 2015. https://thenextweb.com/apple/2015/09/09/genius-annotated-with-genius/.

Wyatt, Sally. "Danger! Metaphors at Work in Economics, Geophysiology, and the Internet." *Science, Technology & Human Values* 29, no. 2 (April 2004): 242–61.

———. "Metaphors in Critical Internet and Digital Media Studies." *New Media & Society* 23, no. 2 (February 2021): 406–16.

Yau, Joanna C., and Stephanie M. Reich. "Buddies, Friends, and Followers: The Evolution of Online Friendships." In *Online Peer Engagement in Adolescence: Positive and Negative Aspects of Online Interaction*, edited by Nejra Van Zalk and Claire P. Monks, 18–34. London: Routledge, 2020. https://doi.org/10.4324/9780429468360-2.

———. "'It's Just a Lot of Work': Adolescents' Self-Presentation Norms and Practices on Facebook and Instagram." *Journal of Research on Adolescence* 29, no. 1 (March 2019): 196–209.

Yin, Rong, and Haosheng Ye. "The Black and White Metaphor Representation of Moral Concepts and Its Influence on Moral Cognition." *Acta Psychologica Sinica* 46, no. 9 (September 2014): 1331–46.

CPSIA information can be obtained
at www.ICGtesting.com
Printed in the USA
BVHW052315030123
655534BV00008B/61